THE SCIENCE OF
STEPHEN KING

THE SCIENCE OF
STEPHEN KING

THE TRUTH BEHIND PENNYWISE, JACK TORRANCE, CARRIE, CUJO, AND MORE ICONIC CHARACTERS FROM THE MASTER OF HORROR

MEG HAFDAHL & KELLY FLORENCE

AUTHORS OF *THE SCIENCE OF MONSTERS* AND *THE SCIENCE OF WOMEN IN HORROR*

Skyhorse Publishing

Skyhorse Publishing books may be purchased in bulk at special discounts for sales promotion, corporate gifts, fund-raising, or educational purposes. Special editions can also be created to specifications. For details, contact the Special Sales Department, Skyhorse Publishing, 307 West 36th Street, 11th Floor, New York, NY 10018 or info@skyhorsepublishing.com.

Skyhorse® and Skyhorse Publishing® are registered trademarks of Skyhorse Publishing, Inc.®, a Delaware corporation.

Visit our website at www.skyhorsepublishing.com.

10 9 8 7 6 5 4

Library of Congress Cataloging-in-Publication Data is available on file.

Cover design by Daniel Brount
Cover photograph by gettyimages

Print ISBN: 978-1-5107-5774-5
Ebook ISBN: 978-1-5107-5775-2

Printed in China

We dedicate this book to all the constant readers
who dare to journey into the darkness.

Contents

Introduction

Every "constant reader" remembers the first time they delved into the delicious horror of Stephen King. For me (Meg), it was finding a paperback of *Carrie* in my parents' stack of books. I asked my mom if she thought I should give it a try. She thought it would be the perfect introduction for me to Mr. King. And, as usual, my mom was right. I was on the cusp of adulthood at age fourteen and, like Carrie, I had a million worries. Unlike Carrie, I couldn't move objects with my mind, or take murderous revenge on my classmates. After reading about the telekinetic teen girl, I was a lifelong constant reader, joining Stephen King on adventures of both unbelievable horror and surprisingly tender heartache. For Kelly, it was watching *The Shining* with her dad in first grade. Having been introduced to horror movies the year before, Stephen King became her new favorite fascination and she sought out his books as she got older.

When we began research for this book, it became clear that we were going to learn about science, folklore, and literary influences. Yet, we ended up coming to understand the author himself more than anything. Join us, fellow constant readers, on a journey into the creative, dynamic world of Stephen King. Leave behind reality for the modest towns of Derry, Castle Rock, and Salem's Lot. Traverse the magical landscape of Mid-World. Because, as you well know, there are other worlds than these.

SECTION ONE

The 1970s

When you think of horror writers, it's impossible not to think of Stephen King. He has scared us, thrilled us, grossed us out, and made us feel terror for nearly fifty years, but his writing breaks the boundaries that are sometimes placed on the genre. As film critic Aja Romano said, "the key to his popularity as a horror novelist, and as a novelist in general, resides not in the darkest moments of his writing, but in his basic belief in humanity's innate goodness."[1] The plots within his works are sometimes otherworldly and sometimes rooted in truth. In this book, we will explore the scientific explanations behind his narratives and the fascinating and sometimes unbelievable horror that exists in our world.

CHAPTER ONE

Carrie

We can't begin talking about the science behind Stephen King's stories until we understand a bit about the man himself. King was born on September 21, 1947, in Portland, Maine, and was raised by a single, working mother. His childhood would hold experiences and memories that inevitably wound their way into becoming parts of his novels, characters, and inspirations for dozens of stories over the years. No one else had as much effect on his writing career, though, as Tabitha.

Tabitha Spruce met Stephen King in the library of the University of Maine, where they were both students, in the 1960s. Both being writers, they attended each other's poetry readings and read each other's work. They married in 1971 and Tabitha encouraged King to write instead of taking a promotion that would leave less time for his craft. This fact led to King's first novel, *Carrie*, being saved from the trash. He had begun writing a story from a woman's perspective, something he took on as a personal challenge after feedback from a reader of his previous work

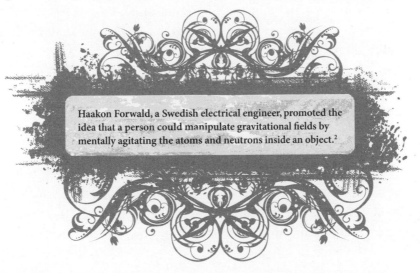

Haakon Forwald, a Swedish electrical engineer, promoted the idea that a person could manipulate gravitational fields by mentally agitating the atoms and neutrons inside an object.[2]

accused him of being scared of women. The story was inspired by an article in *LIFE* magazine that touched on the power of telekinesis. If the power existed, the article purported, then the power would be strongest with adolescent girls. King's first time being in a girl's locker room as a custodian also left an impact on him. Seeing a pad and tampon dispenser seemed almost alien compared to what he was used to.

King wrote three typed pages and then immediately threw them in the trash. The plot seemed to be moving too slowly and he was beginning to doubt his ability to write from a woman's perspective. "I couldn't see wasting two weeks, maybe even a month, creating a novella I didn't like and wouldn't be able to sell," King wrote in his memoir *On Writing* (2000). "So, I threw it away . . . after all, who wanted to read a book about a poor girl with menstrual problems?"[3]

Tabitha retrieved the crumpled pages from the trash and gave him some feedback. In fact, throughout his career, Tabitha is credited with useful, truthful feedback that helps shape characters and mold King's stories. *Carrie* was finished within nine months and sold to Doubleday for a $2,500 advance. Months later, the paperback rights were sold to Signet Books for $400,000. The movie rights sold later, and by 1980 Stephen King was a worldwide bestselling author. As he told the *New York Times*, "the movie made the book and the book made me."[4] The dedication you'll find in every copy of *Carrie* reads "This is for Tabby, who got me into it—and then bailed me out of it."

Carrie was reportedly popular among teen and young adult readers, especially those who could relate to being an outsider. According to Stephen King's website:

> The story is largely about how women find their own channels of power, and what men fear about women and women's sexuality. Carrie White is a sadly mis-used teenager, an example of the sort of person whose spirit is so often broken for good in that pit of man and woman eaters that is your normal suburban high school. But she's also Woman, feeling her powers for the first time and, like Samson, pulling down the temple on everyone in sight at the end of the book.[5]

The themes explored throughout the book are vast; the first is the symbol of blood. "The symbolic function of woman's menstrual blood is of crucial importance in *Carrie*. Blood takes various forms . . . menstrual blood, pig's blood, birth blood, the blood of sin, and the blood of death. It is also blood which flows between mother and daughter and joins them together in their life-and-death struggle."[6] Carrie is unaware that she will begin menstruating. Her mother, Margaret, has intentionally kept this knowledge from her, so when Carrie sees that she is bleeding for the first time, she assumes she's dying. Menstrual blood in particular has been seen throughout history as an abjection or even as supernatural. "The female body and its workings has traditionally been shrouded by misinformation, and historically been a subject that is not supposed to be discussed widely. Reproduction and the menstrual cycle have therefore been viewed as mystical and monstrous."[7]

Margaret White can be described as very religious and her skewed view of the female body and blood are partially based on this. Views of menstruation in religions and cultures vary throughout the world. According to Hippocratic texts from the fifth century BCE, "only female bodies are subject to being overstrained and overfilled due to the excess fluid that accumulates in their inherently soft flesh, and the act of menstruation is a mechanism that releases woman's intrinsic surplus."[8] The Bible claims in Leviticus 15:19 that "whenever a woman has her

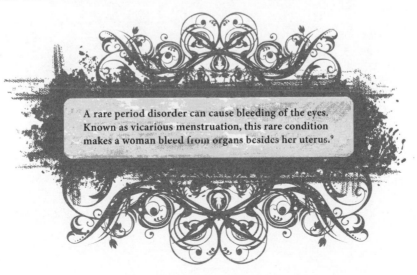

A rare period disorder can cause bleeding of the eyes. Known as vicarious menstruation, this rare condition makes a woman bleed from organs besides her uterus.[9]

menstrual period, she will be ceremonially unclean for seven days." Other cultures viewed menstruating women as powerful or sacred including indigenous people in North America who believed menstrual blood had the power to destroy enemies and the people of ancient Rome who thought a menstruating woman could help crops flourish.

Carrie is told to "plug it up" by her tormentors in the locker room shower that fateful day, but how have women handled menstruation in the past? Assumptions have been made about women using strips of ragged cloth that were rewashed in place of modern-day pads. Ancient tampons were made of papyrus or wooden sticks wrapped with lint. Not until the late 1800s is there documentation of a product being on the market for

The Ishango Bone may have been used as an early period tracker.

women to use during their monthly cycles. The Hoosier sanitary belt was a contraption that was held on around the waist. Washable pads could be purchased that attached to the underwear-like device. The first commercially available tampon was produced in 1929. Just like Carrie, women throughout the centuries have felt a sense of otherness when it comes to menstruation.

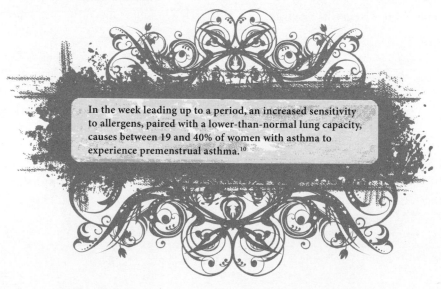

In the week leading up to a period, an increased sensitivity to allergens, paired with a lower-than-normal lung capacity, causes between 19 and 40% of women with asthma to experience premenstrual asthma.[10]

Speaking of menstruation, there is a theory that the first calendar on Earth was created by a woman. The Ishango Bone dates back to between 25,000 to 20,000 BCE and was discovered in 1960 in Zaire. The bone appears to have tally marks on it that document a lunar cycle. American educator and ethnomathematician Claudia Zaslavsky said:

> Now, who but a woman keeping track of her cycles would need a lunar calendar? When I raised this question with a colleague having similar mathematical interests, he suggested that early agriculturalists might have kept such records. However, he was quick to add that women were probably the first agriculturalists. They discovered cultivation while the men were out hunting. So, whichever way you look at it, women were undoubtedly the first mathematicians![11]

This bone, then, could be considered the world's first period tracker! Is it true that periods sync to the lunar calendar in some way? There are theories, particularly in Wiccan and nature-based faith communities, that women can sync their menstrual cycles to a full moon and gain restorative powers. A year-long scientific study, though, found that there is no link between lunar phases and the menstrual cycle.[12]

Another prominent theme in *Carrie* is the effect that Margaret's parenting has on her daughter. An ultra-devout Christian, Margaret White expects her daughter to be pious, chaste, and obedient. What strain does this type of existence put on children? According to a 2008 study, children who grew up in very religious households felt a sense of duty to follow in their parents' footsteps when it came to religious beliefs and practices.[13] Richard Dawkins, a biologist from Oxford, believes that growing up in a religious household could be compared to child abuse. He claims, in relation to the sexual abuse suffered by some children in the Catholic church, that "the damage was arguably less than the long-term psychological damage inflicted by bringing the child up Catholic in the first place."[14] Not all scientists agree, though. A study featured in the *American Journal of Epidemiology* found that religion is linked to better overall health and well-being. They concluded that those who attended religious ceremonies regularly were 12 percent less likely to have depressive symptoms and 33 percent less likely to have sexually transmitted infections. The study also found that 18 percent of self-identified religious people reported high levels of happiness and those who meditated or prayed were 38 percent more likely to volunteer in their communities.[15]

Religion wasn't the only factor in how Carrie was being raised by Margaret White. Research has shown how a mother's lack of affection and focus on authoritarian-style parenting could cause her child to internalize their feelings, ultimately causing anxiety and depression in the child. This authoritarian style of parenting is defined by low levels of warmth and high levels of control, discipline, and punishment.[16] This absolutely describes the White parent-child dynamic. The ideal style of parenting, according to experts, is an authoritative approach. This includes a high level of warmth; nurturing and communication, with a high level of control; setting clear expectations and following through with fair discipline. Even though Margaret may have been raised in a household and culture that influenced her parenting style, this doesn't mean that she couldn't have changed her behaviors for the benefit of her daughter. Culture can explain behavior but doesn't excuse it. In an ideal world, Margaret would have sought out resources for help in balancing her parenting style to help Carrie succeed and gain more self-control.

Another theme in *Carrie* is the psychology of bullying and conformity in high school. Studies have found that bullying in high school is more likely to occur in classrooms and situations in which the norm has been established to support the bullying behaviors.[17] Teens spend most of the day with their peers and will tend to engage in bullying rather than intervene when an incident occurs. Chris Hargensen, as the main perpetrator of Carrie's torment, has a lot of influence over the other girls in the novel. They willingly go along with her actions and are rewarded for their conformity. A 2016 study found that adolescents will often try to avoid negative evaluations from others by keeping their opinions and behaviors in conformity with others.[18]

What sort of long-lasting effects do those who are bullied, or bully, suffer? A study that looked at the psychology of those who were victims and perpetrators of bullying found that:

Kids who had been victims only (who never bullied others) had greater risk for depressive disorders, anxiety disorders, generalized anxiety, panic disorder, and agoraphobia as adults. But worse off were kids who were both bully victims and bullies—they experienced all types of depressive and anxiety disorders, and suffered most severely from suicidal thoughts, depressive disorders, generalized anxiety, and panic disorder, compared with the other groups of participants. In fact, about 25 percent of these participants said they had suicidal thoughts as young adults, and about 38 percent had panic disorder.[19]

If the characters in the novel had survived, they would ultimately have lived with the effects from their actions.

The novel ends with Carrie dying from the wound inflicted by her mother, the ultimate bully. "Blood was always the root of it, and only blood can expiate it."[20] Carrie was powerless but became powerful. She harnessed the strength of her telekinesis to get revenge on those who had wronged her. Although King isn't completely pleased with the novel, it seems an appropriate first venture into publishing. Sometimes the least expected person can come out and surprise everyone in the end.

CHAPTER TWO

The Shining

Since Stephen King came into the literary scene in the 1970s, he has inspired numerous fellow authors. A new generation is finding his writing, thus creating their own brand of horror with odes to King and his impressive body of work. So, it begs to reason that King himself has found inspiration in the authors who came before him. Not surprisingly, he has been quite vocal about the importance of reading if one wants to become a writer. "If you don't have time to read, you don't have the time (or the tools) to write. Simple as that."[1] He further explains how finding inspiration is a good thing. "You may find yourself adopting a style you find particularly exciting, and there's nothing wrong with that. When I read Ray Bradbury as a kid, I wrote like Ray Bradbury—everything green and wondrous and seen through a lens smeared with the grease of nostalgia."

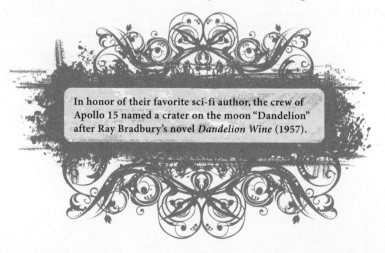

In honor of their favorite sci-fi author, the crew of Apollo 15 named a crater on the moon "Dandelion" after Ray Bradbury's novel *Dandelion Wine* (1957).

It is, in fact, a Ray Bradbury story that first motivated King to form the first buds of what would become *The Shining*. Bradbury's science-fiction short story "The Veldt" published in the 1951 anthology *Illustrated Man* centers on the Hadley family who live in a future where their house does everything for them, including providing a sort of virtual reality that creates your dreams. This concept, of someone's dreams coming true, was at the forefront of King's creation of the novel. King started the story, centering on a psychic boy in an amusement park, in 1972, but was inspired to change the setting when he and Tabitha visited the historic Stanley Hotel in Estes Park, Colorado. One night, King walked the empty hotel as his wife slept. Later, he recounted a strange nightmare:

> That night I dreamed of my three-year-old son running through the corridors, looking back over his shoulder, eyes wide, screaming. He was being chased by a fire-hose. I woke up with a tremendous jerk, sweating all over, within an inch of falling out of bed. I got up, lit a cigarette, sat in a chair looking out the window at the Rockies, and by the time the cigarette was done, I had the bones of the book firmly set in my mind."[2]

It's amazing to think that *The Shining*, such an indelible mark on modern horror, would've never existed if King hadn't visited the Stanley Hotel and been struck with a bad dream! In our book *The Science of Monsters* (2019), we researched the effect of isolation on humans in relation to the Torrance family's long winter in the fictional Overlook Hotel. In fact, there are many fascinating avenues to discover in a novel of ghosts, psychics, and the fracturing of a seemingly all-American family. At the heart of the novel is patriarch Jack, who, like many of King's characters, is a writer and teacher. While the Overlook Hotel may be a catalyst for murder, it is clear that Jack not only has a history of abusing alcohol but also abusing his family. This taboo subject is dealt with subtly

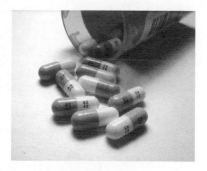

It's estimated that 1.6 percent of the adult US population has borderline personality disorder, but that number may be as high as 5.9 percent and nearly 75 percent of people diagnosed with it are women.

in the novel, as well as in the 1980 Stanley Kubrick film adaptation. In his article "The Shining: The Hellish World of the Tyrannical Patriarch," Jacob Derin delves into both Freudian and Jungian theories to further understand the character of Jack. From a Freudian perspective:

> Classic psychoanalytic thinking focused quite heavily on the pathologies born in the cradle of dysfunctional families. Within this context, The Overlook Hotel's claustrophobic environment represents a dysfunctional household dominated by a dictatorial patriarch who holds sway over a helpless mother and child. Seen in this light, the setting draws us into the insulation and increasingly unhealthy psychological landscape of that family.[3]

And from the archetypal, Jungian perspective, Derin goes as far to say that this "tyrannical patriarch" exists on a larger scale, as countries are often led with such abuse of power.

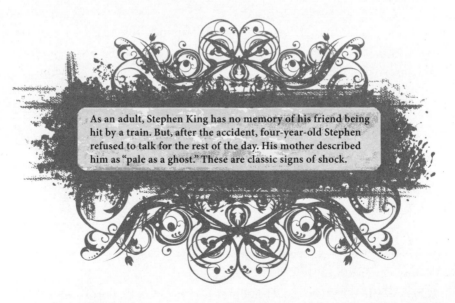

As an adult, Stephen King has no memory of his friend being hit by a train. But, after the accident, four-year-old Stephen refused to talk for the rest of the day. His mother described him as "pale as a ghost." These are classic signs of shock.

However the reader interprets Jack's place in the narrative, there is no doubt he is a flawed human with dysfunctions that are at play long before he arrives at the Overlook. In order to gain a better perspective on the reality of alcohol abuse, a key component to Jack and his bizarre visits to bartender Grady, we spoke to crisis therapist Sara Melendez.

Meg: **"What is your professional background?"**

Sara Melendez: "I have a BA in sociology and an MA in psychology. I also am an Internationally Certified Substance Abuse Counselor."

Kelly: **"Have there been times you have been overwhelmed or emotionally disturbed by your work? How do you deal with the heavy nature of what you do?"**

Sara Melendez: "Yes, I have worked as a crisis therapist for the past fourteen years and have seen thousands of clients in all kinds of disturbing situations—too many to count. I deal with the heavy nature of what I do by understanding that everyone has a journey, and sometimes we don't understand why they are destined to endure so much pain, but it is not my journey, so I have to respect where they're at and the whole process of life itself."

Meg: **"Stephen King has been open about his struggle with multiple addictions, including during the time he wrote *The Shining*. Have you found that creative expression has helped with those grappling with addiction? Can creating art or focusing on hobbies and passions be a part of addiction therapy?"**

Sara Melendez: "In hearing thousands of stories of people in crisis, I have found that some of their stories fall into patterns. That being said, I have yet to meet an individual who suffers a severe mental illness, or a severe substance issue (or both), that hasn't suffered some type of childhood trauma. In order to cope, some follow creative pursuits and when under the influence are able to channel the pain of their childhood trauma into creativity. I've read several times that Stephen King witnessed his friend being hit and killed by a train when they were children. He returned home and had forgotten what happened by the time he got home, according to reports. I believe that event was not resolved at the time he wrote *The Shining*. Mr. King bases some of his stories around childhood trauma like *The Stand* (which also involved a train), *Rose Madder* (1995), *It* (Beverly was a classic victim of childhood sexual trauma by her father). I think that, yes, focusing on hobbies and passions can be part of addictions therapy; however, my belief is that when Mr. King wrote some of his stories in blackout mode (as I've read) he wasn't necessarily in a therapeutic situation but was more likely processing his childhood trauma in his own way."

Kelly: "Wow! That is fascinating, and so true that a lot of his work focuses on dysfunction in childhood, as exhibited by the character of Danny Torrance."

Meg: "I like how authors, or any creators, have the ability to process their trauma through art!"

Alcohol-related fatalities are the third leading cause of preventable death in the United States, following the first, tobacco, and second, obesity.

Kelly: "**In *The Shining*, Jack Torrance is an alcoholic. Soon, the reader finds out that, although he is intelligent and talented, he has lost his job at a prep school. This was due to his drinking which caused him to erupt at his students. He also uncharacteristically broke his son Danny's arm. Can you speak to this Jekyll- and Hyde-like nature in those with addiction? Do you feel this is an accurate portrayal?**"

Sara Melendez: "In some situations, yes. It depends on the substance. I've found that people high on methamphetamine and blackout drunk can be very labile—one second they're reasonable and the next second they're dangerous. People on the influence of other drugs, like heroin, marijuana, and opiates are not as extreme."

Meg: **"As *The Shining* progresses, we witness the conflict in Wendy Torrance's life as the spouse of an alcoholic. What are some of the common issues and traumas that you have come to expect from loved ones in a similar position as Wendy?"**

Sara Melendez: "With every person who is an addict comes an enabler or enabler(s). Wendy was an enabler by tolerating his behavior and staying with him even after he faced consequences due to his addiction (losing his job), and covered up his abuse by not reporting it to authorities after he abused Danny. Without enablers, addicts could not survive long as addicts."

Kelly: **"From your place of expertise, in which King novels or stories were you impressed with his representation of characters either besieged with addiction or mental illness?"**

Sara Melendez: "I was impressed with two characters the most— Beverly in *It* and Rose in *Rose Madder*. Both were not addicts or mentally ill; however, they were victims of close family who had addictions and mental illness. Beverly had an abusive father who was sexually attracted to her, and had difficulty adjusting to her transition to womanhood. It would be safe to say he was a sexual addict as a father attracted to his own biological daughter and struggling to contain his impulses and therefore abusing her. Rose in *Rose Madder* had an abusive, alcoholic husband. What stood out to me the most in this book was that once she left her husband, started a new life, and found a new relationship that wasn't abusive, she became abusive herself and acted out to her new significant other who wasn't abusive at all. In my work with victims of domestic violence, I have found this to be true across the board, although no one really talks about it. Unless a victim of domestic violence works through their issues after they leave the relationship, they are very likely to act out violently toward their new partner, even if the new partner is not violent or abusive. It's like they have all this internal rage. Mr. King touched on that, and it was the first time I've seen it in print, although I've witnessed it in working at domestic violence shelters and with victims of DV."

Kelly: "I'll have to re-read *Rose Madder* with that new perspective!"

Meg: **"Aside from Stephen King, when you see or read fictional accounts of those with addiction or mental illness, do you feel they are depicted fairly? Have there been times you were disappointed with their portrayal?"**

Sara Melendez: "Yes, very much so. Oftentimes the media will vilify a character and label them as bipolar or schizophrenic. They are portrayed as killers, or dangerous, or menaces to society. While there are of course people with either of these diagnoses who break the law, there are so many that are perfectly functional adults. Because I've worked in the trenches throughout my career with people who are bipolar or schizophrenic, I've found that they are amongst us at work, in the community, our neighbor, and sometimes our family. Most of the time as long as they take their medications, stay away from alcohol and drugs, and have support, they are not these crazed insane characters that the media portrays. In fact, those diagnosed with borderline personality disorder do far more damage to families and individuals than others more seemingly 'dangerous' diagnoses."

Kelly: **"What is your favorite Stephen King novel or story and why?"**

Sara Melendez: "Hmm . . . so hard to narrow down, and there are so many! I think *Carrie* really captivated me because I was only a teen and a very sheltered one at that, raised in a dysfunctional household in a small town with a mom who didn't have a grip on reality and a high degree of religiosity. I wished I had superpowers so I could escape. Carrie was my tortured superhero! That was the one story that resonated with me, in a strange way, I guess."

Meg: "What a great example of how fiction can be such a healing and vital part of growing up!"

Thanks to Sara Melendez we were able to stand back and see Jack, Wendy, and Danny Torrance from her perspective as a crisis counselor. This is a testament to the true humanity, whether sweet or depraved, that Stephen King evokes in his novels and stories. While Jack Torrance ultimately succumbs to the evil at the Overlook, the balance of good and evil within him remains the truly compelling aspect of the modern horror masterpiece *The Shining*.

CHAPTER THREE

Salem's Lot

One of my (Meg's) most delightful college courses was a literature class called Tales of Terror. It was naturally popular, in no small part because of its enticing name. As a lifelong horror reader, I made sure to pounce on it once registration was open. We read a number of memorable books and short stories, including Bram Stoker's *Dracula* (1897) and Shirley Jackson's *The Possibility of Evil* (1965). What really tickled me was the chance to deconstruct and discuss a Stephen King novel. As a literature major, I was used to my fellow students maligning King, one very loudly sharing her disdain for the author in our Romantic Literature course. Many had agreed with her analysis, nodding their heads in unison at her insistence that Stephen King was *not* "literary."

I had remained silent, seething underneath.

In Tales of Terror, we were assigned King's 1975 vampire novel *Salem's Lot*. I had already read it as a teenager, but I was more than ready to dive back into the idyllic yet terrifying town of Jerusalem's Lot. In fact, it was because of this novel and the studies of Tales of Terror that I came to understand the term "rural gothic" which has informed much of my fictional work.

American Gothic fiction is a subgenre of Gothic fiction, first developed in Europe with the writings of novelists like Harold Walpole and Ann Radcliffe. American poet and short story writer Edgar Allan Poe popularized the genre, with a macabre atmosphere that nearly always led to death. As Americans took the spooky reins of gothic literature, they branched out into several subtopics. Nathaniel Hawthorne and Washington Irving both wrote upon the dark side of the puritanical, colonial era of America, while authors like William Faulkner are known as purveyors of Southern American Gothic, in which they meditate on the crumbling infrastructure of the southern United States.

The aspects of rural gothic fiction are rather broad, in that they can occur in any place, as long as the setting is far from the bustle of modern, technologically advanced America. What defines rural gothic is not only the subtle aspects that make small-town USA different than the urban city centers, it is also the pervading sense of horror that simmers underneath, hiding beneath the edifice of small-town values. *Salem's Lot* is the perfect gateway into this nuanced subgenre as Stephen King himself explained in a radio interview. "In a way it is my favorite story, mostly because of what it says about small towns. They are kind of a dying organism right now. The story seems sort of down-home to me. I have a special cold spot in my heart for it!"[1]

The town of Jerusalem's Lot, much like King's fictional Maine towns, Derry and Castle Rock, exists in more than one of his works. While it appears first in *Salem's Lot*, it has been revisited in short stories like "One For the Road" (1977) as well as in the last three novels of the Dark Tower Series (2003, 2004). Part of Jerusalem Lot's appeal to readers is its complicated religious history in which a mysterious sect of Puritans mysteriously vanished from the town in the 1700s. This, of course, sounds like the perfect colonial gothic plot for Hawthorne or Irving. But it is the vampires in the 1970s that fold constant readers into the fictional town as they discover along with characters Ben Mears and Susan Norton that beyond the charming buildings and quirky residents, Salem's Lot has a vicious, evil heart.

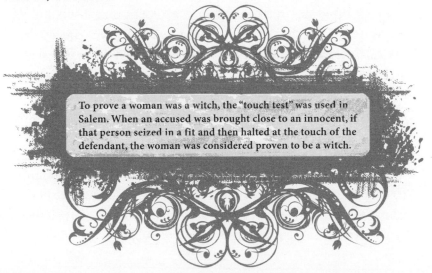

To prove a woman was a witch, the "touch test" was used in Salem. When an accused was brought close to an innocent, if that person seized in a fit and then halted at the touch of the defendant, the woman was considered proven to be a witch.

It was in the real town of Salem, Massachusetts, that true horror unraveled in the 1690s. Many have learned about the Salem Witch Trials, an unsettling piece of New England history which exposed the settlers' very archaic beliefs in witchcraft. Innocent people (mostly women and children) died because of both lack of scientific understanding as well as mass hysteria, or group think, in which beliefs are intensified within a community. Lesser known is a similar panic that echoed the witch trials. Nearly two hundred years after supposed "witches" were hanged by their neighbors and loved ones, a vampire panic overtook another sleepy community in New England.

In 1990, children playing near a gravel mine in Griswold, Connecticut, found a haunting discovery. In order to convince his mother that the skeletal remains were indeed authentic, one boy brought home a skull as proof. The macabre burial site caught the attention of archaeologist Nick Bellantoni, who soon discovered that the bodies had been interred in the early nineteenth century, based on the decay of the skeletons, as well as their meager wooden coffins. Yet, as described in the Smithsonian's "The Great New England Vampire Panic" there was a peculiarity to one grave that intrigued Bellantoni:

Scraping away soil with flat-edged shovels, and then brushes and bamboo picks, the archaeologist and his team worked through several feet of Earth before reaching the top of the crypt. When Bellantoni lifted the first of the large, flat rocks that formed the roof, he uncovered the remains of a red-painted coffin and a pair of skeletal feet. They lay, he remembers, "in perfect anatomical position." But when he raised the next stone, Bellantoni saw that the rest of the individual "had been completely . . . rearranged." The skeleton had been beheaded; skull and thigh bones rested atop the ribs and vertebrae. "It looked like a skull-and-crossbones motif, a Jolly Roger. I'd never seen anything like it," Bellantoni recalls. Subsequent analysis showed that the beheading, along with other injuries, including rib fractures, occurred roughly five years after death. Somebody had also smashed the coffin. The other skeletons in the gravel hillside were packaged for reburial, but not "J. B.," as the 50ish male skeleton from the 1830s came to be called, because of the initials spelled out in brass tacks on his coffin lid. He was shipped to the National Museum of Health and Medicine in Washington, D.C., for further study.[2]

Curious about why this body was so obviously mistreated post-burial, and convinced it wasn't the doing of a typical grave robber, Bellantoni sought the expertise of Michael Bell, New England folklorist and author of *Food for the Dead: On the Trail of New England's Vampires* (2011). Bell had a chilling answer, rooted in the history of nearby Jewett City, Rhode Island, bordering the farm community of Griswold. Based on his research, Bell had come to understand that a panic had spread across the bucolic edifice of the area. Much like in *Salem's Lot*, the townspeople and farmers believed there was an invading, supernatural evil. Why this shared belief swept New England, and as Bell asserts, as far west as Minnesota, is a complex question. What is an important connection, is that the exhumations of bodies, in which loved ones went as far as decapitating or burning the hearts of their loved ones in order to prevent further "vampirism," is that tuberculosis gripped the region. Consumption, as it was called at the time, was an extremely painful

A cartoon that appeared in the *Boston Daily Globe* in 1896.

and drawn out death that frightened people in both urban and rural settings. Caused by a bacterium known as Mycobacterium tuberculosis, consumption "may infect any part of the body, but most commonly occurs in the lungs (known as pulmonary tuberculosis). General signs and symptoms include fever, chills, night sweats, loss of appetite, weight loss, and fatigue."[3] It is transmitted by an active patient spitting, coughing, or sneezing into the air.

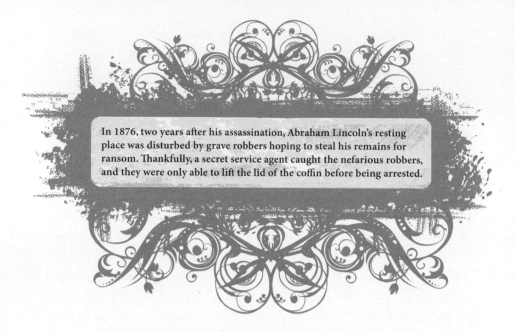

In 1876, two years after his assassination, Abraham Lincoln's resting place was disturbed by grave robbers hoping to steal his remains for ransom. Thankfully, a secret service agent caught the nefarious robbers, and they were only able to lift the lid of the coffin before being arrested.

While consumption has nearly been eradicated from America, it is presently a real threat in developing countries, as nearly 50 percent of the deaths occur in India, China, Indonesia, the Philippines, and Pakistan. At the time of the vampire panic, consumption was sweeping across the world, in a time of only developing knowledge of science and disease. Robert Koch, a German microbiologist, identified the bacteria in 1882, yet it took years for the public to fully embrace the true cause of the disease. As described on the website for the Connecticut Historical Society, blame on consumption's deadly grip shifted depending on the sufferer's background:

> The wealthy blamed it on heredity, as an issue of one's constitution, and so took trips to warm climates for a "cure." For literary types (think Emerson or Thoreau), the culprit was the stress of modern life and their own genius. They sought refuge in nature. The middle class saw the cause as overstimulation from an urban life, including heavy studying, and working in an office. For people living in rural areas, the causes were spiritual and often related to vampirism.[4]

Because of these beliefs in the supernatural, desperation to save loved ones took a macabre turn. Such as in the case of the Brown family, who lived in the farm country of Vermont. The Browns were ravaged by the disease. First, the matriarch Mary Eliza died in 1882. Next, twenty-year-old daughter and sister Mary Olive succumbed to the same disease. Several years later, brother Edwin, formally the heartiest of the family, became ill with consumption. It was ten years after the loss of her mother and sister that Lena Brown began to show signs of the disease. After Lena died, father George Brown was desperate to keep his only living child from following her to the grave. When community members began to imply that perhaps something sinister was at play, George felt no choice but to believe them. With no clue to the realities of bacteria, George came to believe that his dead wife and daughters were rising from their graves and feasting on innocent Edwin. This made sense to him, as consumption's symptoms presented in a sort of "drain" in which the victim slowly lost blood in their cheeks, weight, and strength, until they were a shell of their former selves.

Since there are nine hundred calories per liter of blood, and five liters in a human, a vampire would have to drain one person a day in order to survive.

Neighbors and friends, probably concerned for their own safety, convinced George to allow them to exhume the three women and check for blood in their hearts. This, they believed, would be proof of vampirism. On St. Patrick's Day 1892, George gave the okay. He understandably did not attend the exhumation. "After nearly a decade, Lena's sister and mother were barely more than bones. Lena, though, had been dead only a few months, and it was wintertime. The body was in a fairly well-preserved state," the correspondent later wrote. "The heart and liver were removed, and in cutting open the heart, clotted and decomposed blood was found." During this impromptu autopsy, the doctor again emphasized that Lena's lungs "showed diffuse tuberculous germs."[5] Even more troubling than digging up the deceased, the villagers agreed that the best way to treat Edwin's consumption was to use an inexplicable medicine. They burned Lena's heart and liver, and then fed the ashes to her brother. Not unsurprisingly, this did not cure Edwin, and he died in less than two months.

While it's difficult to understand the beliefs of George Brown and his fellow rural Vermonters today, we can empathize with his great desire to protect his family and community at that time. In *Salem's Lot*, this poignant protectiveness of a small town is on full display. Ben Mears, who spent part of his childhood in the close-knit village, is willing to risk his life in order to save his friends and neighbors from the vampire, Barlow. In fact, Susan Norton, the novel's female protagonist, does just that, dying to save others. Her death, in my opinion, marks one of the most memorable and gut-wrenching deaths in all of King's fiction.

While vampires drain the innocent of their lives, in true rural gothic fashion, Ben must traverse the literal underbelly of Jerusalem's Lot, within the darkened basements and root cellars, in order to slay the beasts. I have to wonder, as I think back on my literature class, Tales of Terror, if what makes *Salem's Lot* so successfully terrifying is that at a time, not so long ago, the people of New England, the world in fact, had to face a true vampire; a vicious and indiscriminate disease.

CHAPTER FOUR

Rage

In 1977, mere months after the publication of *The Shining,* an author named Richard Bachman arrived on the literary scene. Bachman's debut, *Rage*, later compiled in the 1985 collection *The Bachman Books*, was a starkly realistic account of a teenager unhinged. It would be years before it was revealed to the reading public that Stephen King and Richard Bachman were two psyches intertwined, or, perhaps more accurately, one and the same. In the introduction to *The Bachman Books*, Stephen King described the beginning of *Rage*, including its original title. "*Getting It On* was begun in 1966, when I was a senior in high school. I later found it moldering away in an old box in the cellar of the house where I'd grown up—this rediscovery was in 1970, and I finished the novel in 1971." On the official Stephen King website, King answered not only why he adopted the pseudonym for *Rage*'s release, but also where he came up with the moniker:

> The thing is, one book is all most writers want to produce or can produce in the course of a year and some of them only publish a book every two years. Ed McBain is another novelist who publishes multiple books in some years and his original name was Evan Hunter. That's the name he's always published under and he adopted the pen name of Ed McBain for the same reason I adopted Richard Bachman and that was what made it possible for me to do two books in one year. I just did them under different names and eventually the public got wise to this because you can change your name but you can't really disguise your style. The name Richard Bachman actually came from when they called me and said we're ready to go to press with this novel, what name shall we put on it? And I hadn't really thought about that. Well, I

had, but the original name—Gus Pillsbury—had gotten out on the grapevine and I really didn't like it that much anyway, so they said they needed it right away and there was a novel by Richard Stark on my desk so I used the name Richard and that's kind of funny because Richard Stark is in itself a pen name for Donald Westlake and what was playing on the record player was "You Ain't Seen Nothin' Yet" by Bachman Turner Overdrive, so I put the two of them together and came up with Richard Bachman.[1]

Written in first person, *Rage* chronicles the mental deterioration of Charlie Decker, a high school student who has lived through an abusive childhood, thus he is drawn to hateful speech and violence. The novel crescendos when after expulsion by the principal, Charlie chooses to take his misdirected revenge out on his teachers and fellow students. After setting his locker on fire, Charlie, armed with a pistol he had kept at school, shoots and kills both his algebra and history teachers.

The MacDonald Triad, proposed in 1963 by psychiatrist J. M. MacDonald, points to three actions that tend to indicate later violent tendencies: arson, cruelty to animals, and enuresis.

He then keeps his fellow students hostage in a classroom with the power of his gun, reveling in a level of control he has never felt before. That is, until the police force their way in after a four-way stand-off, maiming Charlie and taking him into custody.

While the plot of Rage is unsettling in its harrowing realism of today's school-shooting crisis, it was less culturally relevant in the late 1970s. While school violence has existed since the inception of education, statistics show that the death toll and number of incidents have risen in recent decades. In the study "Historical Examination of United States Intentional Mass School Shootings in the 20th and 21st Centuries: Implications for Students, Schools, and Society," researchers used the definition of mass murder (four or more intentional homicides in one event) as criteria for mapping the rise of mass school shootings in America. For this study, they also included injuries toward the total number. In the 1960s, when Stephen King began writing Rage, there were zero incidents that met this criteria. In the 1970s, in which Rage was published, there was only one mass incident which led to the death of two people. Subsequently, in the 1980s there is the first visual rise on the study's graph. Seven incidents met the mass murder requirement, resulting in the death of twelve. In the 1990s, the incidents nearly doubled to thirteen in which thirty-six people lost their lives. There is a drop in deaths in the 2000s, down to fourteen, and then a dramatic rise in the 2010s, to a staggering fifty-one. Remember, these are mass incidents, and do not include school shootings in which less than four were injured or died.

As referenced in the study, there is also a rise in adolescents rather than adults as the perpetrators, and a possible theory as to why: "Another alarming trend is that the overwhelming majority of 21st century shooters were adolescents, suggesting that it is now easier for adolescents to access guns and adolescents are more frequently suffering from mental illness or limited conflict resolution skills."[2] They also point out that in less than twenty years we have been inflicted with a stunning amount of gun violence in our institutions of learning: "To date, the 21st century shootings have resulted in sixty-six deaths as opposed to fifty-five for the entirety of the 20th century." The school-shooting epidemic is at the center of today's media and politics.

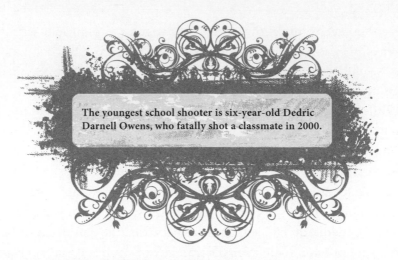

The youngest school shooter is six-year-old Dedric Darnell Owens, who fatally shot a classmate in 2000.

There have been many fingers pointed toward gun control, mental health reform, and school security. In their article "Protecting Students from Gun Violence: Does 'target hardening' do more harm than good?" for *Education Digest*, Bryan R. Warnick and Ryan Kapa suggest:

Educators should also think about how the school climate and culture contribute to the possibility of school shootings and work to change those factors. Reading detailed accounts of school shootings provides clues about what schools could be doing differently. In the early 1990s, the sociologist Katherine Newman led a team of researchers in a study of school shootings since 1970. Their report shone a light on the perennial social competition among teens in the school environment, which Newman termed the "status tournament of adolescence." Some school practices intensify this competition. Think of the prominence of sports in American schools, with the tryouts, rankings, and sorting that go along with it. Think, too, of the teenage fixation on popularity and the common practice of anointing "kings" and "queens" at proms and homecoming dances. School shooters often report feeling like the losers of these status tournaments, and this disappointment sometimes turns to anger against the school environment, as was apparently so in the shootings at Columbine High School in Colorado (1999), East Carter

High School in Kentucky (1993), and Westside Middle School in Arkansas (1998). Instead of fostering competition, schools might look for ways to increase students' sense of belonging.[3]

As the reality of school shootings came into media focus, particularly with the watershed massacre at Columbine High School in 1999, in which thirteen people perished, Stephen King felt a responsibility to censor *Rage*. Today, *Rage* is no longer in print, as reprints of *The Bachman Books* now only contain *The Long Walk, Roadwork*, and *The Running Man*. In fact, much like how Mark David Chapman suggested he was inspired by J. D. Salinger's *Catcher in the Rye* (1951) to murder John Lennon, four school shooters pointed to *Rage* as inspiration for their acts. This included fourteen-year-old Michael Carneal, who in 1997 killed three fellow students at Heath High School in West Paducah, Kentucky. Eerily, there was a copy of *Rage* in Carneal's locker. It was this shooting that prompted King to make certain that *Rage* would no longer be readily available, as he explained in his 2013 nonfiction essay *Guns*.

> I suppose if it had been written today, and some high school English teacher had seen it, he would have rushed the manuscript to the guidance counselor and I would have found myself in therapy posthaste. But 1965 was a different world, one where you didn't have to take off your shoes before boarding a plane and there were no metal detectors at the entrances to high schools.[4]

He further asserts that although he didn't believe *Rage* alone was the cause of the violence, he felt a need to take it off the shelves. "I pulled it because in my judgment it might be hurting people, and that made it the responsible thing to do." This leads us to the question of the true influence of media on violence. This was a hot-button issue when Dylan Klebold and Eric Harris, the shooters at Columbine, were found to be fans of violent video games and "dark" music. Desperate to find the source of such inexplicable tragedy, people pointed to music, video games, films, and books as the reason for the rise in school shootings. In his research study "The School Shooting/Violent Video Game Link:

Causal Relationship or Moral Panic?" Christopher Ferguson compiled data from over fifty sources, coming to the conclusion that:

> There simply is no quality evidence for the predictive value of violent game exposure as a risk factor for school shootings. Indeed, the risk of false positives is significant, even when considered in light with other variables (the inclusion of even one or two "universal variables," that is, variables that are near universally true for the population of interest, give the illusion of multiple risk factors when considered in combination). Even if the focus is on "incessant" interest in violent games, most elders (teachers, parents, psychologists, etc.), as unfamiliar with game culture as most are, simply lack the perspective to evaluate what constitutes "incessant" interest, and what is developmentally normal or even healthy.[5]

Ferguson maintains that it is difficult to find any causation, or to appropriately profile school shooters, as they are such a small sliver of the population and often they are killed by gunfire or suicide. A 2002 study, cited within Ferguson's research by the Secret Service, further highlighted that violent video games cannot be scientifically linked to mass school shooters, saying that only 59 percent of perpetrators demonstrated "some interest" in violent media of any kind (as compared to Griffiths & Hunt's 1995 results suggesting that over 90 percent of nonviolent males play violent video games alone) including in their own writings. For video games, the figure was even lower—only 12 percent.

2018 was a record-breaking year for the industry, with total video game sales exceeding $43.4 billion.

While there is no study on the effects of music on school shooters, there are numerous general studies on the link between music and violence. In 2003, fifty-nine college students participated in a study and separated into two groups in which they listened to music with violent and nonviolent lyrics. In order to keep the study's efficacy, both songs were sung by the metal band Tool. The participants were then tasked with filling out sentences about their current emotions, including rating how they felt on a "hostility scale." This test along with several others led the researchers to conclude that "the violent content of rock songs can increase feelings of hostility when compared with similar but nonviolent rock music. It is important to note that this 'violent lyrics effect' occurred in the absence of any provocation."[6] It is later said, after numerous tests with hundreds of more subjects, that "repeated exposure to violent lyrics may contribute to the development of an aggressive personality."

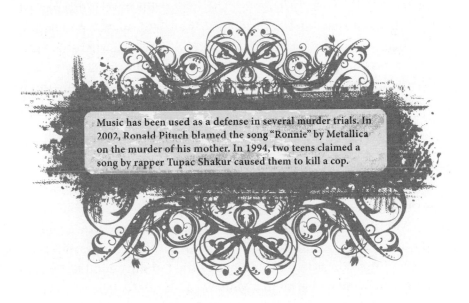

Music has been used as a defense in several murder trials. In 2002, Ronald Pituch blamed the song "Ronnie" by Metallica on the murder of his mother. In 1994, two teens claimed a song by rapper Tupac Shakur caused them to kill a cop.

The reality of school shootings is clearly more complicated than any one influence or personality trait. At the end of *Rage*, Charlie Decker is committed to a mental institution. He has no answer for his sudden, violent rage. Written before its time, *Rage* is a haunting look at the warped and troubled mind of a teenager provoked to perpetrate senseless murder. It is a novel and character that are perhaps closer to our modern, true horror than we'd care to admit.

CHAPTER FIVE

The Stand

As I (Kelly) write this in the spring of 2020, the coronavirus is prominent in all of the current news headlines. A new, deadly outbreak of a virus is wreaking havoc across the globe with panic setting in and death tolls rising. If it sounds familiar, it's because it's eerily similar to the plot of Stephen King's 1978 novel *The Stand*. The virus in the novel is nicknamed Captain Trips and ends up wiping out nearly all of the human population on Earth. By comparison, COVID-19 currently has somewhere between a 3–7 percent mortality rate. Hopefully we won't go down the same route as the civilization did in that story, but maybe we can learn something about humanity from it. The idea to write this epic story had long been in Stephen King's mind:

> For a long time, ten years, at least, I had wanted to write a fantasy epic like *The Lord of the Rings* (1954), only with an American setting. I just couldn't figure out how to do it. Then, slowly after my wife and kids and I moved to Boulder, Colorado, I saw a *60 Minutes* (1968–) segment on CBW (chemical-biological warfare). I never forgot the gruesome footage of the test mice shuddering, convulsing, and dying, all in twenty seconds or less. That got me remembering a chemical spill in Utah that killed a bunch of sheep (these were canisters on their way to some burial ground; they fell off the truck and ruptured). I remembered a news reporter saying, "If the winds had been blowing the other way, there was Salt Lake City."[1]

What was the chemical spill Stephen King was referring to? The Dugway sheep incident took place in 1968 in Utah when six thousand sheep were killed due to chemical and biological weapons testing. The

Dugway Proving Ground had been testing nerve agents in the days prior. One test involved the firing of a chemical artillery shell, another the burning of 160 gallons of nerve agent in an open-air pit, and a jet aircraft spraying nerve agent in a target area.[2] Although the sheep were grazing twenty-seven miles west of the testing area, over six thousand died as a result of the nerve agent. Answers about the incident aren't definitive, but some people believe a malfunction of a nozzle could explain what happened.

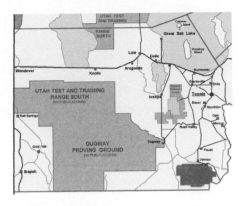

The Dugway Proving Ground is a US Army facility established to test biological and chemical weapons.

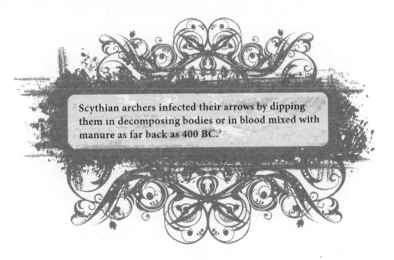

Scythian archers infected their arrows by dipping them in decomposing bodies or in blood mixed with manure as far back as 400 BC.[3]

Biological warfare is another theme that inspired Stephen King to write *The Stand*. Although it seems like a modern-day invention, biological warfare is not a new tactic. One of the first recorded instances of this type of warfare took place in 1155 when Emperor Barbarossa poisoned a water well with human bodies in Tortona, Italy. In 1346, Mongols catapulted bodies of plague victims over the city walls of Caffa in the Crimean Peninsula. The Spanish mixed wine with blood of leprosy patients to sell to their French foes in Italy in 1495 while the Polish fired saliva they took from rabid dogs at their enemies in 1650. In the United States, the British distributed blankets from smallpox patients to indigenous people in 1763 to try to get them to spread the disease unknowingly. The effectiveness of this last example is unknown, but experts speculate it wasn't successful because smallpox is spread more efficiently through the respiratory system. In 1797, Napoleon flooded the plains around Mantua, Italy, to try to spread malaria and in 1863, Confederates sold clothing from yellow fever and smallpox patients to Union troops in the United States to spread disease.

More recent biological weapons include the development and deployment of anthrax and sarin gas. Anthrax inhalation is often fatal. Initial signs and symptoms of inhalation of anthrax include flu-like symptoms, shortness of breath, nausea, coughing up blood, fever, and possible meningitis. In 2001, twenty-two people got anthrax through letters that were sent through the mail and five of them died. Sarin gas was initially developed in 1938 in Germany as a pesticide. According to Dr. Lewis Nelson from Rutgers New Jersey Medical School:

> Sarin targets an enzyme within the body's neuromuscular junctions, where nerves meet muscles. Usually, this enzyme deactivates the nerve-signaling molecule acetylcholine. But sarin stops this deactivation by blocking the enzyme. Without the enzyme to switch it off, acetylcholine will repeatedly stimulate nerve cell receptors. This can lead acetylcholine to build up in the muscles, cause excessive twitching and then result in paralysis. If the muscles that control breathing become paralyzed, the person can die.[4]

Sarin gas was most recently used in the 2017 attacks in Syria resulting in the deaths of at least eighty-six people, including twenty-eight children.

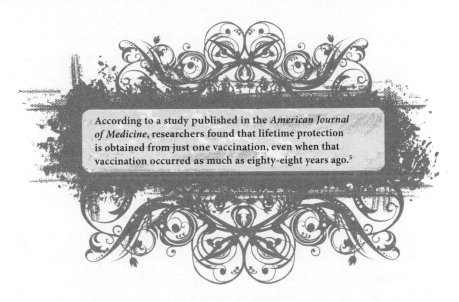

According to a study published in the *American Journal of Medicine*, researchers found that lifetime protection is obtained from just one vaccination, even when that vaccination occurred as much as eighty-eight years ago.[5]

The pandemic that wipes out nearly all of the human population in *The Stand* is far deadlier than any to hit the globe so far. What exactly is a pandemic? The word pandemic comes from the Greek *pandemos* meaning "pertaining to all people." Outbreaks of diseases that cross international borders are considered pandemics including cholera, bubonic plague, smallpox, and influenza. Smallpox, which has killed between three hundred to five hundred million people in its twelve-thousand-year existence, is one of the deadliest. Smallpox is a virus that begins with fever, aches and pains, and sometimes vomiting. A rash will appear that starts as small red spots on the tongue and in the mouth. Next, the rash will appear on the skin and will usually spread to all parts of the body within twenty-four hours. Thanks to the smallpox vaccination, the World Health Assembly declared the disease eradicated in 1980.

The Black Death was a plague that ravaged Europe, Africa, and Asia from 1346 to 1353. With an estimated death toll between seventy-five and two hundred million people, this bubonic plague devastated entire continents. Symptoms include fever and chills, headache, muscle pain, general weakness, and seizures. The disease is caused by a bacterial strain called Yersinia pestis which is found in animals throughout the world and is usually transmitted to humans through fleas. The risk of plague is highest in areas that have poor sanitation, overcrowding, and a large population of rodents. This disease is not eradicated but the total number of cases per year is down to one to two thousand.

The flu pandemic of 1918 infected over five hundred million people in the world and it's estimated that around fifty million people died from it. The strange and scary thing about this particular flu was that it killed those who were normally very healthy and hearty while it is the very old and very young who are normally the most in danger. In order to try to reduce the spread of the disease, many businesses and schools were closed for periods of time. People were discouraged from gathering in large groups and from spitting on the sidewalk. In Minnesota, a Dr. Delmore in Roseau County advised eating squirrel soup! The symptoms of this particular strain of flu could come on quickly and some people died within the same day that they began experiencing them. In 2008, it was discovered why the disease was so deadly. "A group of three genes enabled the virus to weaken a victim's bronchial tubes and lungs and clear the way for bacterial pneumonia."[6]

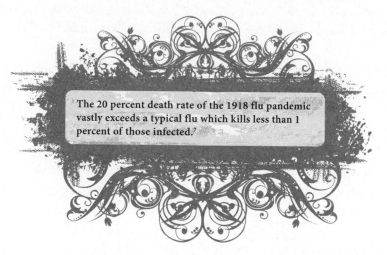

The 20 percent death rate of the 1918 flu pandemic vastly exceeds a typical flu which kills less than 1 percent of those infected.[7]

HIV/AIDS is considered a pandemic and has killed over thirty-six million people since 1981. The symptoms include rapid weight loss, recurring fever or profuse night sweats, extreme and unexplained tiredness, prolonged swelling of the lymph glands in the armpits, groin, or neck, diarrhea that lasts for more than a week, sores of the mouth, anus, or genitals, and pneumonia. New treatments and education have helped lower the death rates from this horrific disease and have helped those who live with it have a better quality of life.

While in the midst of the coronavirus panic, it's interesting to note that the medical community is aware of how big a problem a modern-day pandemic can be. According to an article in *Medical News Today*, "It can take months or years for a vaccine to become available, because pandemic viruses are novel agents. Medical facilities would be overwhelmed, and there could be shortages of personnel to provide vital community services, due to both the demand and illness."[8] Only time will tell what this current health crisis will hold for the future of our world but in the meantime, stay well, dear reader, don't forget to wash your hands, and curl up with a good book.

CHAPTER SIX

Nightshift

Stephen King released his first short story collection in 1978 with *Nightshift*. Nine of the twenty stories featured in the book had appeared in publications throughout the 1970s. With his popularity rising, students began approaching King about adapting his short stories into plays or short films. He approved the policy of the "Dollar Baby," which allows the creator to adapt his work for the price of $1. Numerous works have been created over the years through this program, giving filmmakers a unique opportunity to work with the Master of Horror's fiction.

Gray Matter

One of the short stories in *Nightshift* was adapted for the screen on the television show *Creepshow* in 2019. "Gray Matter," which was first published in 1973, takes place in the evening during a snowstorm. Richie Grenadine is a recluse who hasn't appeared in public for a while. He sends his son to the local convenience store to pick up his beer but

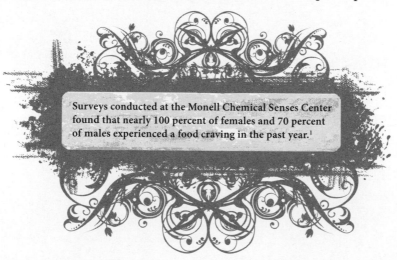

Surveys conducted at the Monell Chemical Senses Center found that nearly 100 percent of females and 70 percent of males experienced a food craving in the past year.[1]

this evening, Richie's son seems worried. The group of men at the store decide to deliver the beer to Richie themselves so they can get an idea of what is happening in the home. They discover that the beer Richie has been drinking has turned him into a horrible beast.

Richie is craving warm beer in "Gray Matter." (There must be something wrong with him!) What is the science behind cravings? Scientists explain that an area of our brain which is responsible for memory, pleasure, and reward also controls our urge for certain foods. "The hunger loop operates out of the hypothalamus at the base of the brain. The reward system, on the other hand, is located in the center of the brain. It involves many regions of the brain, such as the ventral tegmental area, the nucleus accumbens, and the prefrontal cortex."[2] We may crave junk food, sugary treats, or a salty snack. It all depends on the person and the mood. Some cravings may be due to an imbalance of hormones, such as leptin and serotonin, or can be due to endorphins that are released into the body after someone has eaten, which mirrors an addiction. Emotional eating and cravings are common for people, and pregnant women do indeed experience especially strong cravings. Is it true that we crave things that our body needs? For example, chocolate cravings could be blamed on low magnesium levels, whereas cravings for meat or cheese could be seen as a sign of low iron or calcium levels. Scientists don't agree on whether our bodies alert us to food needs. It's recommended to get enough sleep, drink enough water, and eat a balanced, healthy diet to reduce cravings. I'm not sure if that advice would have saved poor Richie in "Gray Matter" but it's worth a shot!

There are over seven thousand beer breweries in the United States.

The inciting incident that led to Richie's demise was a can of beer that was a bit off. To understand more about the scientific process of brewing beer we spoke to a beer brewing enthusiast, and Meg's husband, Luke Hafdahl, to learn more about his process.

Kelly: **"How did you get interested in brewing beer?"**

Luke Hafdahl: "I suspect like most others, it starts with just drinking and enjoying beer! But I was intrigued by the creative process. I have a background in chemistry and biology, so I really became hooked when I began to see how much we harness basic science to brew beer and it really tapped into my enjoyment of experimentation. Brewing truly is a wonderful hybrid of creativity and science."

Meg: **"What sort of training did you need in order to start?"**

Luke Hafdahl: "Like so many things in life, if you are pursuing it as a hobby, it is a very easy thing to teach yourself. The main brewing process is simple and could be likened to teaching yourself how to bake. However, if you truly want to take it seriously as a career, you really need to be trained. There are many programs across the country, some of them four-year programs. It truly is a scientific discipline that requires a firm understanding of chemistry, physics, and microbiology. For home brewing, you can get by with a cursory grasp of these principles, but to be considered a brew master, you need training to become a brewing scientist."

Kelly: **"What is the process for brewing and bottling beer?"**

Luke Hafdahl: "In its most basic form, brewing involves three things: grain (typically barley), hops (for bittering and flavor), and yeast (they are the workers that make the alcohol). That's it. You boil grains and hops to release sugars that the yeast eat for food. It's their 'waste' product that is the alcohol that we want (this is called fermentation).

Meg: **"In Stephen King's story 'Gray Matter' the character of Richie drinks a can of beer that's 'gone bad.' Although we discover in the story the origin of the problem is otherworldly, what really could cause a can or bottle of beer to go bad?"**

Luke Hafdahl: "The single biggest thing that one worries about when brewing is contamination of the beer. When we brew, we are very thoughtful about the strain of yeast that we use because that is how we produce a predictable flavor (in addition to making alcohol, yeast will make other by-products that can add flavors). These yeast have cultured under controlled conditions so that when you purchase them, you know exactly what you are getting. However, yeast live everywhere: on our skin, on our countertops, literally everywhere. And these 'wild' yeast (as well as some bacteria!) would love nothing more than to get into the beer we are making and eat up all the sugars themselves. Thus, if you do not pay close attention to sanitizing your equipment (killing the wild organisms) when brewing, they can take over the fermentation process and make some awful flavors. Some describe the flavors like 'wet cardboard,' 'metallic,' and 'musky.'"

Meg: "Oh yes, some of my favorites!"

Kelly: **"Do you remember your first experience reading a Stephen King book or watching a movie based on one of his works? Has it had any lasting impact on you?"**

Luke Hafdahl: "My first exposure to Stephen King was at the fourth-grade book fair when I bought one of his books called *Eye of the Dragon*, which literally is probably the only Stephen King book nobody is talking about adapting into a TV show or movie. I heard a lot about Stephen King, and even though this was probably one of the most unusual entry points into his works, I was compelled by his storytelling and I went on to read a lot of his books from there. I really love fantasy (*Lord of the Rings*, *Game of Thrones*, etc.) and I suspect this book had a lot do with that as it really captured my imagination and exposed me to the limitless potential of fantasy."

Thank you to Luke Hafdahl for letting us in on his process for brewing beer (and for letting us sample some of his brews)!

Richie is described as more of a fungus than a human by a certain point in "Gray Matter." Is it possible for fungi to grow on people? There are over 1.5 million types of fungi in the world and 300 of them can cause illness in humans. Some of these diseases include athlete's foot, ringworm, aspergillosis, histoplasmosis, and coccidioidomycosis. We all have fungi living in our bodies, too, and they can be healthy for us. The give-and-take among bacteria, viruses, fungi, and each person's specific biology likely influences our health. When out of balance, oral fungi can cause thrush and interactions between fungi and bacteria in the gut can aggravate the body's autoimmune response in Crohn's disease.[3]

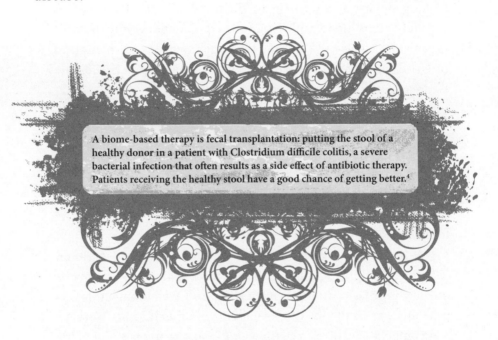

A biome-based therapy is fecal transplantation: putting the stool of a healthy donor in a patient with Clostridium difficile colitis, a severe bacterial infection that often results as a side effect of antibiotic therapy. Patients receiving the healthy stool have a good chance of getting better.[4]

Richie is dividing and multiplying at the end of the story and will eventually conquer humanity. Are there examples in nature of things that divide in this way? Molds, yeasts, and mushrooms are all able to use fragmentation in order to reproduce. In the animal kingdom, sponges, coral colonies, annelids, and flatworms reproduce by this method. This is bad news if a giant fungus is ever trying to take over the world! We don't find out the plan of attack by the end of the story and that makes it all the more terrifying.

Quitters, Inc.

More people in the United States are addicted to nicotine than to any other drug. Quitting smoking isn't easy for most people but for Dick Morrison in the short story "Quitters, Inc.," it becomes a necessity. Approached one day by an old acquaintance, Dick is given a business card for a place that can help people with nicotine addiction called "Quitters, Inc." The mysterious outfit recruits its clients by word of mouth and has a purported 98 percent success rate of getting people to stop smoking. Dick isn't given any specifics about the program but is intrigued enough to stop in one day and see what it is all about. What he discovers after he signs a contract is that if he doesn't quit smoking, his wife and son will suffer the consequences for his actions. Someone will always be watching him and if he smokes another cigarette, he will be punished. The barbaric ramifications would begin with his wife getting electric shocks while he watched and escalate to his son being physically beaten.

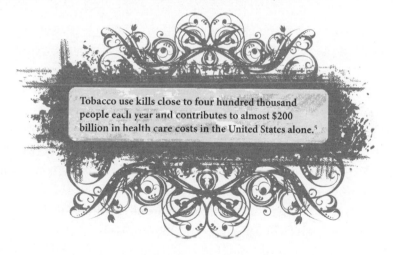

Tobacco use kills close to four hundred thousand people each year and contributes to almost $200 billion in health care costs in the United States alone.[5]

Torture and fear are used as a motivational tool in this story, but is this an effective strategy? Studies have shown that when negative reinforcement is used with animals, it produces backward or freezing motions in them. When this type of motivational tool is used with people in a work environment, it leads to lower productivity, an unpleasant work environment, and an increase in absenteeism. Why is this so? Our brains are wired to respond to these outside stimuli. We have rewards to ensure repeated behavior and rewards to ensure that a behavior is terminated. Our most accessible memories tend to be the ones that have an extreme positive or negative reward experience attached to them.

Torture is not a new concept when motivating people. The ancient Romans and Greeks incorporated torture into their judicial systems and legal proceedings. Torture during the Middle Ages was often a public spectacle used to induce fear into the spectators. The torture of those who were presumed to be witches was common during the early modern period and it wasn't until the late eighteenth and early nineteenth centuries that torture was legally abolished.

Dick Morrison only succumbs to one cigarette during his "treatment" and pays the price. This motivates him to quit smoking for good and he becomes part of the 98 percent success rate. How difficult is it to quit smoking? According to the Centers for Disease Control and Prevention, nearly 68 percent of smokers want to quit. Formal programs have a higher success rate for cessation of the habit compared to those who try to go it alone. People who complete a treatment program are successful 20 to 40 percent of the time while those who try to quit on their own are only successful in 3 to 5 percent of cases.[6]

Why is it so difficult to give up nicotine? Nicotine is considered to be just as addictive as heroin, cocaine, or alcohol. The body's response to the drug is to release a kick of adrenaline. This stimulates several parts of the body including the brain which releases dopamine in the pleasure portion. Nicotine causes people to feel less stress, feel more relaxed, and experience a higher level of concentration. But not all the effects are positive. Nicotine causes an increase in blood clotting, the forming of plaque in artery walls, and changes in blood pressure.

The benefits of quitting smoking are immediate and robust. Those who quit have a lowered risk for lung cancer, a reduced risk for heart disease and stroke, and reduced respiratory symptoms, such as coughing, wheezing, and shortness of breath.[7] Within days people also report an increase in their sense of smell and taste. Regardless of your method for quitting smoking, it is recommended that you stop. But probably do your research on the company that you sign up with so you don't end up like Dick Morrison!

SECTION TWO
The 1980s

CHAPTER SEVEN

Cujo

Stephen King knew he was an addict in 1975 and struggled with alcohol and other addictions over the next decade. Somehow, he was able to write the novel *Cujo* while in a complete blackout state. He doesn't remember much about the writing process itself, but he does remember the inspiration. In 1977, he took his motorcycle to a mechanic who lived outside of Bridgton, Maine, as King described to be "in the middle of nowhere."

> I took the bike out there, and I just barely made it. And this huge Saint Bernard came out of the barn, growling. Then this guy came out and . . . I was retreating, and wishing that I was not on my motorcycle, when the guy said, "Don't worry. He don't bite." And so I reached out to pet him, and the dog started to go for me. And the guy walked over and said, "Down Gonzo," or whatever the dog's name was and gave him this huge whack on the rump, and the dog yelped and sat down. The guy said, "Gonzo never done that before. I guess he don't like your face." And that became the central situation of the book, mixed with those old "movies of the week," the made-for-television movies that they used to have on ABC. I thought to myself, what if you could have a situation that was an extension of one scene. It would be the ultimate TV movie. There would be one set, there would be one room. You'd never even have to change the camera angle. So, there was one very small place, and it became Donna's Pinto—and everything just flowed from that situation—the big dog and the Pinto.[1]

The novel is written without chapters, akin to a stream-of-consciousness flow. "There's one novel, *Cujo*, that I barely remember writing at all. I don't say that with pride or shame, only with a vague sense of sorrow and loss. I like that book. I wish I could remember enjoying the good parts as I put them down on the page."[2] Some view the novel as one of King's greatest metaphors for addiction. This beast is lurking, trying to kill us, and we may feel completely helpless against it.

There's an estimated seventy-seven million dogs as pets in the United States.[3]

We see dogs as our domesticated best friends but they weren't always viewed in this way. To understand more about how our perception of dogs evolved into what it is today we spoke with former science teacher (and Kelly's dad) Robert Maki about the history of dogs.

Meg: **"The theory of evolution of dogs begins with wolves becoming domesticated. Can you explain how this came about genetically?"**

Robert Maki: "I believe a genetic mutation occurred in the past that led to a more passive wolf. This 'friendlier' wolf must have mated with other wolves with similar genetic traits. Watching a *NOVA* (1974-) special on PBS a while back, I believe they said that three genes are different with the wilder wolves and the friendlier wolves. The friendlier wolves eventually became domesticated dogs."

Kelly: **"Some scientists theorize that humans began domesticating dogs as far back as twenty thousand years ago while others claim that it is more recent than that. When do you believe this domestication occurred?"**

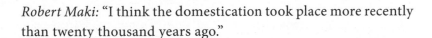

Robert Maki: "I think the domestication took place more recently than twenty thousand years ago."

Kelly: **"Why do you think that?"**

Robert Maki: "Lone wolves, not in a pack, came to campsites with a 'friendly' demeanor looking for food. The lone wolves met and bred with other lone wolves who had similar less aggressive characteristics. These wolves/dogs became helpers for tracking prey, warning of intruders, companions, and protection."

Pure wolves are illegal to keep as a pet and are protected under endangered wildlife species regulation.[4]

Meg: **"What are the theories about how dogs began to interact with early humans? What was their relationship like?"**

Robert Maki: "I believe that early interaction started with food. A wolf probably smelled the cooking venison or rabbit and investigated what the smell was. When the wolf visits didn't result in a dangerous situation, the wolves became less afraid. If early humans gave the wolves food, a bond was beginning."

Meg: **"Have you had dogs as pets in your life? What were some of your favorites?"**

Robert Maki: "All my life I have been around dogs. My first dog was a female cocker spaniel named Cindy. I had her for many years. We slept together and played together. My sheets and blankets were usually full of chew holes."

Meg: "I can relate to that! I've had more than a few blankets destroyed by one of my dogs."

Robert Maki: "I was going to college when Cindy was hit by a car. I still remember burying her in the woods."

Kelly: "That first pet death is the worst! It's like losing a member of the family."

Robert Maki: "I was heartbroken. Most of our dogs were given to us by people who couldn't keep them or we found dogs who had been abandoned, or they found us! My favorite dog in my adult life was Huckleberry, a husky mix. He was very friendly and affectionate to me, although he sometimes scared other people. Every winter he would go cross-country skiing with me for hours at a time. We did everything together. In our pickup, he would sit next to me and put his chin on my shoulder. He was friendly to our cats and spent a lot of time grooming them."

Kelly: "I loved that dog!"

Kelly: **"I grew up knowing who Stephen King was because we watched *The Shining* together when I was in the first grade. When was the first time you remember reading a Stephen King book or watching one of his movie adaptations? What was that experience like?"**

Robert Maki: "I have always liked Stephen King books and movies. *The Shining, Cujo, Carrie, The Green Mile, Christine,* and *Pet Semetary* were some of my favorites. I always enjoyed getting goosebumps on my arms during a scary part of a book or movie."

Meg: **"Do you have a favorite book or movie of his? What do you like about it and what sort of impression did it make on you?"**

Robert Maki: "It is hard to choose, but *Pet Semetary* is one of my favorites, probably due to the number of pets I have had throughout my life. To bury the dead pet and then have it come back to life was one of the scariest scenes because the pet had changed drastically. The father then buried his dead child in the pet cemetery hoping that he would come back to life. To find the result of this situation, you will have to read the book or see the movie!"

Meg: "No spoilers here!"

Although dogs have become man's best friend, there are still a number of dog attacks that occur daily. According to a study from the Centers For Disease Control, almost five million dog bites occur in the United States each year, and eight hundred thousand of those bites result in medical care. Why does a

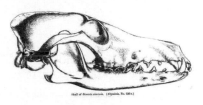
Skull of *Bassin aiervin.* (Alysion. No. 1264.)

Wolves and modern-day dogs are considered the same species.

dog bite a person? A dog may attack due to their reaction to a stressful situation, if they feel scared or stressed, or if they feel they need to bite in self-defense. In the case of Cujo, the dog may bite if they are not feeling well or if they are sick. To help a dog get to know you, it's always recommended to let it sniff your hand before attempting to pet it. Typically, between thirty and fifty people die per year in the United States from a dog attack. To put it in perspective, there is an average of one fatality every two years from shark attacks. Both are rare, but dog attacks are more likely.

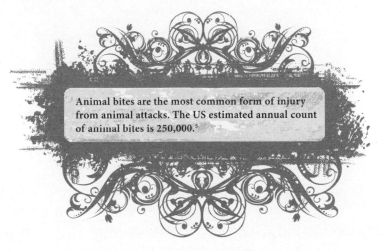

Animal bites are the most common form of injury from animal attacks. The US estimated annual count of animal bites is 250,000.[5]

Donna and Tad Trenton are locked in their car being tormented by the dog Cujo throughout much of the novel. Tad succumbs to the heat in the vehicle and dehydration rather than an attack from the dog itself. How long would it take to get dehydrated in a car? A car sitting in the sun can gain forty degrees of heat in just under an hour. Our bodies will react and begin to sweat in order to cool us down. This sweating will cause dehydration, electrolyte imbalance, and can lead to heat stroke or even death. An average of thirty-seven children die per year in the United States from being left in hot cars.[6] It's recommended that you always park your car in the shade, when possible, or use a sun shield to help prevent extreme temperature increases. Never leave your children or pets unattended in a hot car, of course, and set a reminder on your phone or other device to help you remember your loved ones in the backseat.

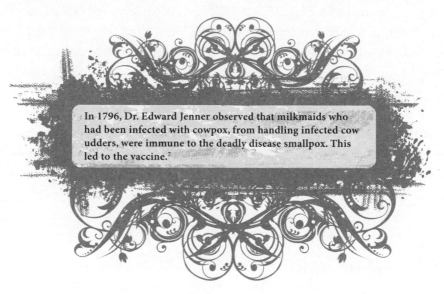

In 1796, Dr. Edward Jenner observed that milkmaids who had been infected with cowpox, from handling infected cow udders, were immune to the deadly disease smallpox. This led to the vaccine.[7]

The reason for Cujo's behavior comes down to his owner missing the dog's rabies vaccination. What is the history behind animal vaccinations? The science behind animal vaccinations is a relatively recent field. French chemist Louis Pasteur developed a vaccine for chicken cholera in 1879. It was discovered by accident when Pasteur realized that a forgotten cholera sample lost its ability to transmit the disease. He hypothesized that this low dose of the disease would allow the chickens' bodies to build a defense against the stronger strain. He was correct and saved chickens from dying due to this lethal disease. Pasteur developed another vaccine for anthrax of sheep and cattle in 1881. He tested his rabies vaccine on animals in 1884, and within a year, the vaccine's success prompted its use on humans bitten by suspected rabid dogs.[8] Donna is able to defeat Cujo by the end of the novel just as Stephen King was able to defeat his addictions. It's not always easy, the path may not be clear, but it's important to fight to save your life.

CHAPTER EIGHT
Pet Sematary

If you've had a pet die, you know how sad the process can be. Stephen King experienced this firsthand in 1979 when his daughter's cat was struck by a passing truck. The idea for the novel *Pet Sematary* was born out of this experience. In the woods behind his house, local children had created an informal pet cemetery where he buried the cat.

> I can remember crossing the road, and thinking that the cat had been killed in the road and (I thought) what if a kid died in that road? And we had had this experience with Owen running toward the road, where I had just grabbed him and pulled him back. And the two things just came together: on one side of this two-lane highway was the idea of what if the cat came back, and on the other side of the highway was what if the kid came back, so that when I reached the other side, I had been galvanized by the idea, but not in any melodramatic way. I knew immediately that it was a novel.[1]

That night he dreamt of a reanimated corpse and that spurred him to thinking about death and burial customs.

Is it possible for corpses to be reanimated? Although reanimation has been talked about in myth and legend for centuries, it wasn't until the 1700s that experiments were documented. Lazzaro Spallanzani, a Catholic priest and natural history professor, believed that the dead could be reanimated using water. He noticed that microscopic life, which appeared dead, seemed to come back to life when water was added to it. Spallanzani never observed reanimated life but did end up observing white blood cells. In 1794, the Royal Humane Society of London carried out experiments on corpses in the hopes to alleviate the public's fears about premature burial. This process included pouring liquor down

the corpse's throat, blowing smoke up the rectum, and massaging the dead body to attempt to awaken it. In the 1800s, physicist Giovanni Aldini became famous for his demonstrations of "reanimating" human and animal corpses. The process involved stimulating them with electrical shocks that would cause the corpses to convulse as though they were alive. In the 1930s Robert E. Cornish, a biologist at the University of California Berkeley, began experiments to bring dead dogs back to life. He used a contraption that would swing the corpse around, as if riding a seesaw, while he administered oxygen, adrenaline, liver extract, and

In the Bible, Jesus raised Lazarus from the dead after four days of burial.

anticoagulants. Cornish was able to bring two dogs back to life who reportedly both lived for several months afterward. Animal experiments of this type continued in Russia and the United States throughout the subsequent decades.

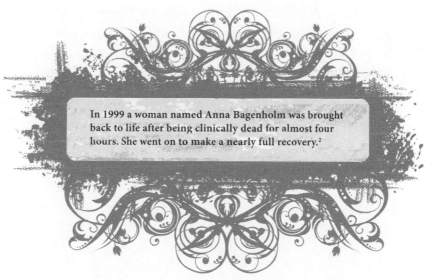

In 1999 a woman named Anna Bagenholm was brought back to life after being clinically dead for almost four hours. She went on to make a nearly full recovery.[2]

Why is it so difficult to bring living things back to life? Scientifically speaking, when our bodies die they begin the process of cell death.

All of our cells are covered with a thin membrane that essentially protects it from its surroundings and filters out molecules that are not necessary to its survival. When a cell approaches death, this membrane becomes thin and the cell will either be absorbed by surrounding specialized maintenance cells, it will basically eat itself, or the cell membrane will rupture, its contents spectacularly spewed into the surrounding tissue. Once any of these three things happen, there is no going back and the cell's death is final. When this final cellular death occurs, reanimation becomes impossible.[3]

When someone dies, the lack of oxygen to their brain causes cell damage or even cell death. This is why when many people are revived or brought back to life, they are essentially brain-dead. This may explain why zombies in the media act the way they do! It also gives credence to the way the cat and son behave after being revived in the novel *Pet Sematary*.

A character in the novel that played a bigger role in the film version is Zelda Goldman, Rachel's sister. She suffered from spinal meningitis and Rachel still feels a combination of guilt and resentment over her condition. What is spinal meningitis? According to Cedars Sinai Hospital, spinal meningitis is "an infection of the fluid and membranes around the brain and spinal cord. Once infection starts, it can spread rapidly through the body."[4] Symptoms may include a blank, staring expression,

Every year over one million people worldwide are affected by meningitis.[5]

a dislike of being touched or handled, a high-pitched, moaning cry, an arching back, and a pale, blotchy skin color. All of these symptoms could have been terrifying for Rachel to witness and incredibly painful for Zelda to endure. This trauma, and the trauma of her son's death, had a tremendous impact on Rachel's life.

What is the science behind trauma? Through research, we now know that caring, positive relationships have an effect on how our brains develop as children. The opposite of that type of environment can have a negative effect on the brain not only in childhood, but throughout life. The brain learns to adapt to its environment so when someone is in a state of constant fear or panic, their brain will react to situations in that frame of mind. The good news is that the brain can also be taught how to reverse these negative effects. Having close, consistent, positive relationships is the key to overcoming trauma at any age. One key component for adults is to use mindfulness techniques in order to be present in their current situation. By being mindful and logical about their feelings, people who have experienced trauma can retrain their brains.

A theme in *Pet Sematary* is the idea of sacred burial grounds and the power they might possess. Americans have been fascinated with indigenous burial lands for centuries. Philip Freneau wrote in his 1787 poem "The Indian Burying Ground,"

> Thou, stranger, that shalt come this way,
> No fraud upon the dead commit—
> Observe the swelling turf, and say
> They do not lie, but here they sit.[6]

Around the time of the writing of *Pet Sematary*, there were real land disputes at play in the state of Maine that Stephen King was hearing about. These disputes put into the minds of all landowners whether they truly own the land they have paid for. This guilt or observance that comes from being aware of indigenous peoples' rights to land plays a part in the horror genre. Those who disrespect the land or steal from rightful owners will be haunted or punished.

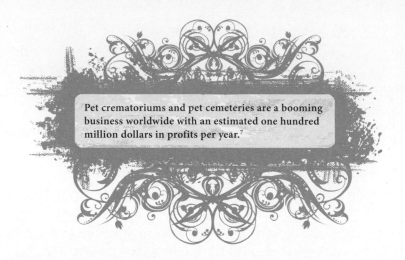

Pet crematoriums and pet cemeteries are a booming business worldwide with an estimated one hundred million dollars in profits per year.[7]

Another Native American theme that is prevalent in the novel is the legend of the Wendigo. We had the opportunity to interview artist Winona Nelson, a member of the Leech Lake Band of Minnesota Chippewa, about the legend.

Kelly: **"In the latest iteration of Stephen King's *Pet Sematary*, the burial ground's powers are rooted in a long mythology, stemming back to the people who inhabited the northeastern seaboard and continental interior, especially the Great Lakes region. Do you know any legends specifically about this region and indigenous people?"**

Winona Nelson: "Yes, I know a bit. My tribe, the Ojibwe, is spread throughout that region. Because it is so far north and the climate so inhospitable, we survived the coming of the Europeans in large numbers and were able to keep a lot of our traditions and stories and our language strong, compared to what happened to tribes to the south and east of our range."

Meg: **"The word *wendigo* translates to 'the evil spirit that devours mankind,' though one German explorer translated it to mean 'cannibal.' How does this definition fit the legend through the Ojibwe story?"**

Winona Nelson: "The Ojibwe language has concepts that can't translate very well to English, because a word can refer to both a literal creature or thing as well as a metaphor or idea. Wendigo stories are sometimes simply monster stories meant to scare or thrill. As a monster, a wendigo is often described as hugely tall, a former human cursed with a desperate hunger that cannot be satiated no matter how much they consume, because when they eat they only grow larger, needing more food. They are so thin they are like skin stretched over a skeleton, and are inhumanly strong and cannot be killed without first being turned back into a human by forcing them to drink hot tallow."

Kelly: "That sounds terrifying!"

Winona Nelson: "Wendigo stories come from the very real risk of starvation faced by Ojibwe people. Winters in the north are brutal. Communities would break apart into independent family groups because it is easier to keep a small group alive. In extreme times it wasn't unheard of for someone to turn to cannibalism to survive. This is a major taboo, and was a crime for which the Ojibwe had a death penalty, because once a starving person has turned to cannibalism there is no longer a barrier keeping them from doing it again. In those cases, a wendigo is a real-life monster, a human who has broken the worst taboo and must be executed. Wendigo also refers to all-consuming greed, such as that of the oil industry and capitalism and the profit-over-everything mentality that is killing our planet. So, both of the translations are accurate, yet don't quite cover all the meanings the word has to us."

Kelly: **"According to the Canadian Encyclopedia, 'wendigo legends are essentially cautionary tales about isolation and selfishness, and the importance of community.' Do you think this theme is prevalent in fairy tales and legends we pass down through generations?"**

Winona Nelson: "Yes, although different cultures have different balances of themes in their stories. In general, European fairy tales place more importance on piousness, obedience, treasure, and royalty than the stories of North American tribes, which focus more on individual freedom, respect for all beings, and self-knowledge. If the Europeans had more wendigo stories instead of so many fairy tales ending with rewards of wealth, maybe we wouldn't now be living in such an unsustainable way."

Thank you to Winona Nelson for this eye-opening interview!

"Sometimes dead is better,"[8] Jud Crandell says in *Pet Sematary*. We agree. Although it can be sad to lose a loved one or close pet, don't try to reanimate them. It never seems to turn out the way you'd like.

CHAPTER NINE

Thinner

Many of us have wished that we could eat as much as we want and not gain weight. My husband's family is notorious for having exceptional metabolism and has been encouraged to eat more. I (Kelly) have the opposite problem and could gain five pounds by just looking at a piece of cake! Billy Halleck, the main character in Stephen King's *Thinner*, has a similar problem in the beginning of the book, but through a curse is unable to keep any weight on. Stephen King recalled being heavier himself at one point:

> I used to weigh two hundred and thirty-six pounds, and I smoked heavily. I went to see the doctor and he told me, "Listen, man, your triglycerides are really high. In case you haven't noticed it, you've entered heart attack country." I used that line in the book. He told me that I should quit smoking and lose some weight. I spent a very angry weekend off by myself. I thought about it and how awful they were to make me do all these terrible things to save my life. I did lose the weight, and pretty much quit smoking. Once the weight actually started to come off, I began to realize that I was attached to it, somehow, that I didn't really want to lose it. I began to think about what would happen if somebody started to lose weight and couldn't stop. It was a pretty serious situation at first. Then I remembered all the things I did when I weighed a lot. I had a paranoid conviction that the scales weighed heavy, no matter what. I would refuse to weigh myself, except in the morning, and then after I had taken off all my clothes. It was so existential that the humor crept in after a while.[1]

This journey of weight loss led to the novel *Thinner* being born.

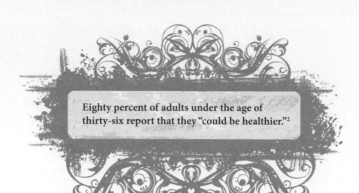

Eighty percent of adults under the age of thirty-six report that they "could be healthier."[2]

Released in 1984, *Thinner* follows the story of what happens to the main character after he commits vehicular homicide one night. Being a lawyer, and with good connections, Billy Halleck faces no legal repercussions from his crime. He faces real consequences, though, when he is cursed by the father of the victim saying a single word to him, "thinner," after his trial. There are numerous examples of "gypsy curses" being used in horror media over the years, often in retribution for an unpaid debt of some kind. This trope is often seen as harmful and demonizing to the Romani people. It is estimated that there are currently over one million Romani Americans living in the United States. Having originated from Northern India, the Romani people were first brought to America as slaves in 1492 and later immigrated during the nineteenth and twentieth centuries. The term "gypsy" is frowned upon by the Romani people as they have been unfairly persecuted for centuries and stereotyped as fortune tellers, kidnappers, and thieves. The truth, of course, is that the Romani people, like everyone, can't fit into a narrow definition. Unfortunately, some stereotypes persist in the media.

There are other examples of "curses" throughout history among various groups of people. One of the most famous curses surrounds the finding of Ötzi the "Ice Man" in 1991. Rainer Henn, a forensic pathologist who examined Ötzi, died in a car accident a year later. Kurt Fritz, the guide who led Henn to Ötzi's body, died in an avalanche shortly thereafter. Helmut Simon, who first discovered the body, died from a fall while hiking in 2004. Dieter Warnecke, who headed the rescue team looking for Simon's body, died of a heart attack just hours after Simon's funeral.[3] These could all be coincidences, of course, but it doesn't stop the human imagination from wondering about another explanation for these deaths.

Susanna Lemke, the woman killed in the novel *Thinner*, is trying to cross the street as a pedestrian when she is struck by Billy's car. How common is it for people to be killed in traffic accidents? According to the Governor's Highway Safety Association, pedestrian fatalities have increased 41 percent from 2008 and

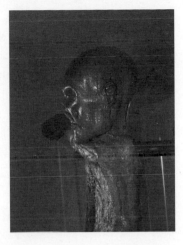

Ötzi the "Ice Man" was found to have sixty-one tattoos and an intestinal parasite!

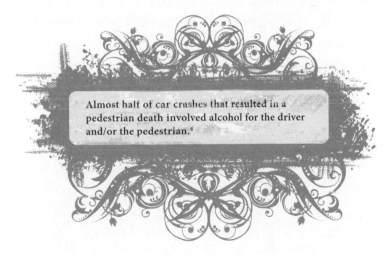

Almost half of car crashes that resulted in a pedestrian death involved alcohol for the driver and/or the pedestrian.[4]

are at an all-time high.[5] That amounts to 6,227 deaths in the year 2018 alone. Some contributing factors include the use of smartphones, by both drivers and pedestrians, and bigger vehicles.

Something that is more common, especially in the United States, is the epidemic of obesity. Since Stephen King wrote *Thinner*, the obesity rate has almost tripled. To be considered obese, a person's body mass index (BMI) has to be over thirty. In 1960, less than 14 percent of Americans were considered obese while in 2019, 40 percent were.[6] When considering the definition of being overweight, the percentage jumps to a staggering 80 percent of the American adult population. How did we get here? The common medical advice is to eat less and exercise more. If the answer is so simple and readily available, it seems like that percentage should go down. A number of factors contribute to people trying to lose weight and not being able to keep it off, including the popularity of fad diets, which can work but are unsustainable and can even biologically change your metabolism.

After the curse, Billy Halleck isn't having trouble with his weight loss anymore. The pounds are flying off and he's able to eat just as much, and even more than he previously did. What are some possible causes for this rapid weight loss, besides curses, that could be explained by science? One example sounds like it's straight out of a horror movie: a foreign creature inhabits your body and feeds off your food supply. Fiction? Hardly. Tapeworms and parasites can absolutely infect humans and sometimes grow up to eighty feet and live as long as thirty years![7] (No, thank you!) Women in the Victorian era were known to swallow tapeworms willingly to try to lose weight and the fad hasn't disappeared. In recent years, people have been ingesting tapeworms in the hope of

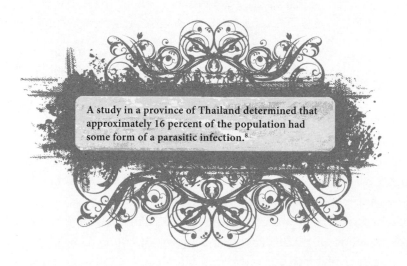

A study in a province of Thailand determined that approximately 16 percent of the population had some form of a parasitic infection.[8]

losing a few pounds. Medical professionals do not recommend this fad diet, however, as it can cause complications including diarrhea, infection, and blockage of the intestine. (Not to mention a creature living inside you!)

Other reasons for rapid weight loss include gastrointestinal diseases such as celiac disease, Crohn's disease, and ulcerative colitis. Those who suffer from these conditions often experience diarrhea and malabsorption of nutrients which causes the sufferer to lose weight. Cancer, Addison's disease, and hyperthyroidism could also be to blame which could include symptoms of fatigue, muscle loss, and increased metabolism. Doctors recommend that anyone experiencing unexplained rapid weight loss make an appointment to explore the possible reasons behind it.

Another curse in the novel *Thinner* is put on the character of Judge Cary Rossington with the word "lizard." He is cursed with scaly skin and even begins to grow a tail in the film version. Do either of these conditions exist in real life? There are several skin conditions that may make the skin appear to have scales including eczema, psoriasis, and ichthyosis. Worldwide, about 20 percent of the adult population suffers from eczema while 3 percent of children have some form of the disease.[9] Certain foods can trigger symptoms, including nuts and dairy as well as environmental factors such as smoke and pollen. Treatments for eczema include moisturizing, using mild soaps, and taking prescription medications. Psoriasis affects more than eight million people in the United States and can be mild, moderate, or severe. For a case to be considered severe, 10 percent or more of the body would be covered with psoriasis. Light therapy and prescription medications are used to treat the condition. People who suffer from ichthyosis, also known as fish scale disease, have a genetic disposition to either shedding their old skin cells too slowly or growing new skin cells too quickly. Ichthyosis affects 1 in 250 people in the United States. Treatments include taking salt baths, moisturizing with special creams, and taking retinoid medications. The skin condition the Judge had in the novel seemingly couldn't be cured, though, and he perished.

The lizard-like tail and clawed hands that Judge Rossington is suffering from are two distinct features some humans are born with.

Vestigial tails grow during the fifth week of the gestation period in the womb and disappear by the eighth week. If the tail doesn't get absorbed or form into the tailbone, a protrusion resembling a tail remains. Since they serve no function, the tails are often removed shortly after birth. The longest known tail on a human belongs to Chandre Oram of India who, due to spina bifida, sports a thirteen-inch tail. Clawlike hands are also a real condition and can occur due to nerve damage, leprosy, or as a birth defect.

Duncan Hopley, the chief of police in *Thinner*, is given the curse of acne for his role in the death of Susanna Lemke. How common is acne? Acne is extremely common and is one of the top three most encountered dermatological conditions. The diagnosis and treatment of acne has existed since ancient Greek and Egyptian times and focused on the four humors of the body, believed to be blood, phlegm, black bile, and yellow bile. Treatments throughout the centuries ranged from natural remedies, such as honey, to changes in diet. Prayers and the balancing of the four humors were also a focus of acne treatment up until the Middle Ages. Topical creams and prescription medications became more common in the twentieth century and most acne is able to be treated using one of these methods.

Ultimately, Billy Halleck is able to reverse the curse by giving it to someone else via a "gypsy pie." He doesn't escape his fate, though, because instead of just passing the curse on to his wife he gave it to his daughter as well. Fittingly, the novel ends with him about to take a bite of the cursed pie and secure his demise.

CHAPTER TEN

It

There are few horror icons who exist in the cultural zeitgeist that surpass their movies and books. They are boogeymen scaring children who know only their iconic images. Michael Myers in his hockey mask, Freddy Kreuger with his knife fingers, and, of course, Pennywise the clown. With his resurgence of popularity thanks to the recent films, *It* (2017) and *It: Chapter Two* (2019), Pennywise has come back with a vengeance, terrifying a new generation.

Published in 1986, and clocking in at an impressive eleven hundred pages, *It* is considered one of King's masterpieces. A novel about both the trauma and deep friendships forged in childhood, *It* resonates with any reader who's ever been a misfit, or simply had a deep-seated fear. The author explained how he approached such an epic novel: "I've written some books and [I've] gotten this reputation as a horror novelist, so IT will be my final exam. I'll bring back all the monsters that I remember from my childhood . . . because the entity that is Pennywise focuses upon whatever that particular child fears the most."[1]

Perhaps it is this flexibility that makes Pennywise so terrifying. For the character of Eddie, plagued by an overprotective mother, he is haunted by a leper covered in disease. For Bev, a girl abused by her father, she is tormented by her own reality, something more threatening than any creepy clown. Yet, when readers remember the most pulse-pounding moments of *It*, it is Pennywise, simply an extension of the hungry creature beneath Derry, that endures.

So, what is it about clowns that cause many humans to feel discomfort, or even fear? When I (Meg) was five, I distinctly remember a tin trash can I had inherited from my big brother. It was decorated with dancing, juggling clowns. One night, as the moon filtered in through my bedroom window, I noticed that these painted clowns were *moving*. In the eerie light they looked as though they were trying to escape their trash can prison. I leapt from my bed, raced to my parents' bedroom, and woke my mother, begging her to remove the suddenly animate clowns from my room. She obliged, though it still took me hours to fall asleep!

Years later, a mother now myself, my youngest has developed an inherent fear of clowns. At about three years old, at the annual Fourth of July parade, he cried at the sight of the Shriners dressed in silly clown costumes. Older now, he still steps back when they come by, visibly shaken by their painted smiles. I'm sure there are many reading this book who have had their own frightful encounter with clowns, whether they have been exposed to Pennywise or not. This is naturally counterintuitive, as clowns were developed to bring smiles, laughter, and joy to both children and adults. So, why does coulrophobia (the fear of clowns) exist? Is the media to blame? Or is there something naturally disturbing about their presence? Ben Radford, author of *Bad Clowns* (2016) posits that "It's misleading to ask when clowns turned bad, for they were never really good. You can no more separate a good clown from a bad clown than a clown from his shadow."[2]

The history of the clown is varied. It is said to have been derived from an archetype known as Zanni, an Italian comic character who was portrayed as foppish and jester-like, in the Commedia dell'Arte. The Zanni was known to act like a drunkard to amuse the rich Italian class.

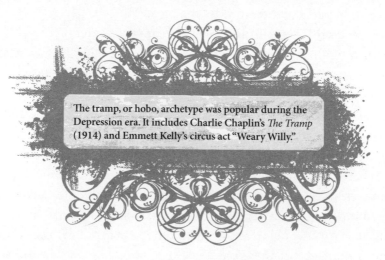

The tramp, or hobo, archetype was popular during the Depression era. It includes Charlie Chaplin's *The Tramp* (1914) and Emmett Kelly's circus act "Weary Willy."

He was also constantly hungry, willing to eat almost anything, which is ironic in the context of *It*, as the creature behind Pennywise has a hankering for innocent children! Those acting as Zanni would often dress in baggy, white clothes, similar to the peasant class at the time. This is reminiscent of the clothes later worn by more modern clowns.

In the early days of Commedia dell'Arte, the Zanni mask used in the theater was a full face mask with a long nose. It developed from here into a half mask covering the upper half of the face only with an extended, long nose. The longer the nose on the mask, the more stupid was the character. The costume of the Zanni character reinforced for the audience the nature of this character. Usually dressed in sacking and hunched over, carrying heavy loads as a porter, with knees in a "bowed" position and feet splayed apart. This was in sharp contrast to the aristocratic characters in the Commedia dell'Arte who always carried themselves with erect deportment. Zanni was always highly animated, waving arms and gesticulating with hands when speaking in a coarse manner with erratic body movements thrown in! Some Zanni characters were known for their acrobatic feats, including handstands and flips.[3]

As comedies developed across the world, different types of the jester, or Zanni, came to be. Today, France is known for their silent, white-faced mimes. While in America, through the popularization of the circus, our tastes changed from the "hobo" or "tramp" character, to the Auguste, meaning "red clown" in the bright wig, red nose, and oversized shoes. This style of clown fully infiltrated American culture with the advent of *The Bozo Show* (1960–2001) as well as the instantly recognizable fast food mascot, Ronald McDonald.

With every joyous clown, intent to evoke smiles, there seems to be a contrasting evil clown. These clowns may dress and paint their faces with the same vivid colors, yet their intentions are far more nefarious. In 1940, Batman villain the Joker was introduced to comic readers across the world. Like Pennywise, the Joker has become an American icon, appearing in numerous TV shows and films. In all his iterations, he retains the Auguste aesthetic, yet is murderous to the core. In 1982, a clown toy terrorized children in the film *Poltergeist*, not long after the apprehension of the real killer clown.

John Wayne Gacy, one of the most notorious serial killers of the twentieth century, was caught in 1978 after a reign of terror that shocked not only his local Chicago suburb, but the world. Convicted of thirty-three murders, Gacy had hidden twenty-six of his victims in the crawl space of his home. His victims were all male, many of them teenagers. At a time when serial murder was rather a new concept, the notion that Gacy was a successful businessman and community volunteer made the reality all the more terrifying. And on top of it all, the images of Gacy as "Pogo the Clown" brought goosebumps. Gacy enjoyed performing for children as Pogo, dressing in the typical Auguste style with his signature triangular eye paint. Along with the Joker and Pennywise, Pogo has joined the pantheon of evil clowns, all the more horrendous, as he did not spring from the imagination, but is the real thing.

In the winter of 2017, before *It* hit theaters, researchers published a study in the *European Journal of Pediatrics*. Their aim was to discover if children in pediatric wards were scared of medical clowns, and if so, why. Medical clowns have long been a hospital tradition, think Robin Williams's wacky character in *Patch Adams* (1998). The researchers used a sample of over one thousand children, and involved the medical clowns in the study. After the clowns entertained the children for about ten minutes, with the typical tricks and prop gags, they filled out a form about how many of the young patients experienced coulrophobia, or a fear of clowns. Children and parents voiced their opinions as well.

Out of the 1160 children, only 14 children experienced coulrophobia of any severity (1.2 percent). The average age of children experiencing coulrophobia was three and a half years (range one to fifteen years). Most of the children who experienced coulrophobia were girls (twelve out of fourteen). Of the fourteen children experiencing coulrophobia, six had it in a severe form (43 percent) responding to the clown's visit with significant fear, crying, and apposition behavior. The rest of the patients exhibited moderate fear (eight children, 57 percent). Based on the a-priori determined criteria, none responded with mild coulrophobia. The anxiety responses observed by the research assistant were variable: crying, anger, standing still, and holding on to the caregiver. Eight out of the fourteen participants in the coulrophobia group responded with crying during the medical clown visit. Out of fourteen participants in the coulrophobia group, twelve were reported as trying to avoid further contact with the medical clown (hiding, staying in the room, etc.)[4]

This led the researchers to conclude that it was a relatively low
number of patients who had actual
coulrophobia. Their reasoning was
that evil clowns were not as prevalent
in the media as they had been in the
1980s. Also, the medical clowns in
the study wore little makeup, which
may have been a contributing factor.
We have to wonder, with the advent
of the two new *It* films, if the study
would have different results now?

In 2016, there were over one hundred suspicious clown sightings in the United States which led to some arrests.

One reason clowns may bring us unease is the concept of the uncanny. German psychiatrist Ernst Jentsch first coined the term, with Sigmund Freud further developing the idea. The uncanny essentially refers to some-thing familiar though mysterious. Clowns are humans, yet their makeup, dress, and behavior are in contrast to what we expect. Freud wrote:

> Frightening things would then constitute the uncanny; and it must be a matter of indifference whether what is uncanny was itself originally frightening or whether it carried some other effect. In the second place, if this is indeed the secret nature of the uncanny we can understand why linguistic has extended das Hemlich ('Homey') into its opposite das Unhemlich; for this uncanny is in reality nothing new or alien, but something that is familiar and old-established in the mind and which has become alienated from it through process of repression.[5]

While this concept may sound esoteric in the writings of Freud, it really comes down to the unnatural feeling we often sense around humanlike robots, dolls, and those pesky clowns.

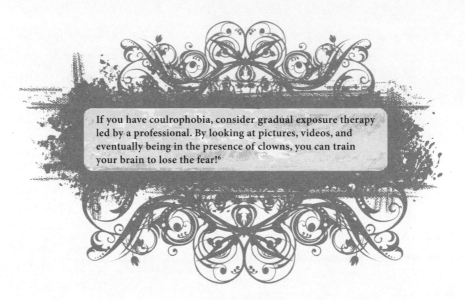

If you have coulrophobia, consider gradual exposure therapy led by a professional. By looking at pictures, videos, and eventually being in the presence of clowns, you can train your brain to lose the fear![6]

The creature living in the sewers of Derry will do anything to scare the children above. They taste better scared! No wonder it plays upon their greatest internal fears to lure them in. Little did it know that the Losers Club would work together to kill it. Pennywise is an extension of the creature, yet he has become the face of *It*, spawning many look-alikes in haunted houses across the world. There was even a surge of clowns terrorizing small-town America in 2016. From Ohio to South Carolina and beyond, reports of clowns stalking the night flooded emergency services. In the documentary *Wrinkles the Clown* (2019), an unidentified man chronicles his quest to scare "misbehaving" children, paid by their parents.

If the presence of a clown makes you uncomfortable, you can thank Stephen King. While the media might not be *all* to blame, it certainly seems to exacerbate our already unnerved reaction to the uncanny. As the author himself says, "Clowns are scary. There's just no way around that. Clowns can be as angry as they want, and that's their right—they're clowns!"[7]

CHAPTER ELEVEN

The Drawing of the Three

If you're familiar with Stephen King's work, then you notice when the number nineteen pops up. In his introduction to *The Drawing of the Three*, King recalls nineteen was the age he was inspired by J. R. R. Tolkien's *The Lord of the Rings* to write an epic of his own. He put it off, but kept the thought in his mind. Years later, while watching Sergio Leone's movie *The Good, the Bad and the Ugly* (1966), King realized he wanted to combine Tolkien's sense of quest and magic set against a majestic Western backdrop.

King's other inspiration for the series included Robert Browning's 1855 poem "Childe Roland to the Dark Tower Came." Several character names are contained within it and the plot even mirrors some of the imagery created by it:

> Thus, I had so long suffer'd, in this quest,
> Heard failure prophesied so oft, been writ
> So many times among "The Band"—to wit,
> The knights who to the Dark Tower's search address'd
> Their steps—that just to fail as they, seem'd best.
> And all the doubt was now—should I be fit?[1]

The poem has been interpreted by many and after Stephen King read it in the late 1960s, it stuck with him.

Browning never says what that tower is, but it's based on an even older tradition about Childe Roland that's lost in antiquity. Nobody knows who wrote it, and nobody knows what the Dark Tower is. So, I started off wondering: *What is this tower? What does it mean?* And I decided that everybody keeps a Dark Tower in their heart that they want to find. They know it's destructive and it will probably mean the end of them, but there's that urge to make it your own or to destroy it, one or the other. So, I thought: *Maybe it's different things to different people, and as I write along I'll find out what it is to Roland.*[2]

This journey to discover the Dark Tower along with Roland is one that has captivated readers for decades and concluded in the final novel of the series, *The Dark Tower VII: The Dark Tower* in 2004.

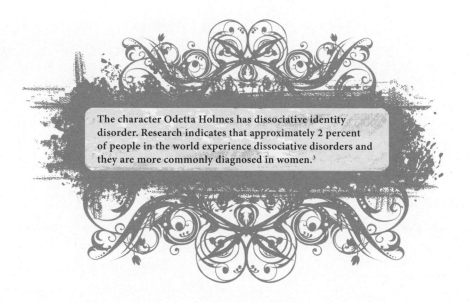

The character Odetta Holmes has dissociative identity disorder. Research indicates that approximately 2 percent of people in the world experience dissociative disorders and they are more commonly diagnosed in women.[3]

One constant of our love for Stephen King stories is the characters that he creates. He is able to deftly describe people so that even if we can't relate to them, we are able to understand them perfectly. The character Eddie is addicted to heroin in *The Drawing of the Three*. What is the science behind addiction? The National Institute on Drug Abuse (NIDA) defines addiction as "a chronic, relapsing brain disease that is characterized by compulsive drug seeking and use, despite harmful consequences."[4] When we are addicted to something, it changes the physiology in our brains and the way they function. Some people use drugs to feel better while others use them out of curiosity. Depending on the person and their ability to exert self-control, addiction may become inevitable.

To understand more about addiction, we interviewed Andrew Memelink, a substance abuse counselor, about his experiences working in this field.

Addiction changes the physiology of the brain.

Kelly: **"A recurring theme in many of Stephen King's works is addiction. A character in *The Drawing of the Three* is addicted to heroin and quits cold turkey. What is your experience helping people overcome an addiction? What is the process like?"**

Andrew Memelink: "I've worked in the field of addiction for fifteen years, fourteen of which have been as a licensed alcohol and drug counselor. I've run outpatient substance abuse groups, residential treatment groups, been a program director at a treatment center, head counselor at a detox facility, and a counselor at a methadone clinic (MAT).

Process varies greatly but could include drying up or quitting using and detoxing either at a hospital or on their own. The next step is to work on their underlying issues through therapy, treatment, or through AA. They may go to a MAT facility to avoid withdrawal and try to avoid relapse. If they get sober and don't deal with their underlying issues, that usually ends up with being sober and miserable (dry drunk)."

Meg: **"Addiction was primarily considered a moral flaw in the past. Do you think people still view addiction this way and how does that affect their treatment?"**

Andrew Memelink: "In the medical community it's still seen fifty-fifty as a moral flaw versus the disease concept of addiction. The disease concept of addiction is becoming more and more prevalent. In the field of addiction and people working in the addiction field, it's widely accepted as a disease (disease concept of addiction). In the community and people who don't appear to understand addiction, the moral flaw concept is very prevalent. I still hear from patients all the time saying 'my family doesn't understand, they keep asking me why I can't just stop.' The stigma of addiction remains a major barrier in all forms of treatment and recovery in all facets of our society. From a provider standpoint I see the stigma of working with addiction as the primary reason why addiction counselors make a fraction of what mental health therapists earn, why we have to do much more paperwork, and have a much more difficult time receiving funding for our patients to attend treatment or participate in treatment services. In the past two months I have been asked by social workers and chaplains at the hospital I work at to do presentations on addiction and the process of recovery. Fifteen years ago it didn't appear anyone outside of addiction wanted to hear anything about addicts or how they could help them."

Meg: "I'm glad to hear that's changing!"

Drug abuse and addiction cost American society more than $740 billion annually in lost workplace productivity, health care expenses, and crime-related costs.[5]

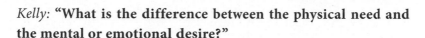

Kelly: **"What is the difference between the physical need and the mental or emotional desire?"**

Andrew Memelink: "The physical part of heroin withdrawal is in the acute stage of withdrawal. More than any other drug, heroin withdrawal is usually the most intense. Common symptoms include full body sweats, frequent vomiting and diarrhea, bone pain (like you got beat up in a gang initiation), and cramping. The mental and emotional side in acute withdrawal is irritation, frustration, depression, and difficulty finding pleasure in anything. Post-acute symptoms are physically minimal but the mental and emotional ones are vast. They include cravings, ongoing difficulty feeling pleasure, struggles dealing with difficult situations, sleep problems, depression, and anxiety. The sleep problems can include very lucid or realistic dreams. It's not uncommon for people in early recovery to report having used in their dreams and waking up feeling high."

Kelly: "That's fascinating! And it must be scary to wake up and believe all the progress you've made has been undone."

Meg: **"What advice would you give to someone trying to overcome an addiction?"**

Andrew Memelink: "It's not going to be easy, it will be hardest and the worst at the beginning. It will get better over time; the key is to keep fighting. Deal with your underlying issues so they don't keep dealing with you. Life can be better than you can imagine at this time."

Kelly: **"Are you familiar with any Stephen King books or movies? Do you have a favorite?**

Andrew Memelink: "*Rita Hayworth and the Shawshank Redemption* (1982) and *Different Seasons* (1982) were phenomenal in general. *Storm of the Century* (1999) also was always a favorite of mine."

Kelly: "Those are all great! Thank you for speaking with us."

How does heroin affect the body? Heroin binds to and activates specific receptors in the brain called mu-opioid receptors (MORs). They release dopamine and the brain's reward centers are activated. When heroin enters the brain, it's converted to morphine and users describe a pleasurable rush feeling. The good feelings are offset by potential complications including sleepiness and slowed breathing, which can lead to possible brain damage and even coma. Prolonged use of heroin can permanently affect the brain. According to The National Institute on Drug Abuse, "studies have shown some deterioration of the brain's white matter due to heroin use, which may affect decision-making abilities, the ability to regulate behavior, and responses to stressful situations."[6] The process of detoxing from heroin can be quite painful and uncomfortable with symptoms ranging from muscle aches and nausea to shaking and depression. These symptoms tend to be strongest within twenty-four to forty-eight hours after using the drug and can last up to several months.

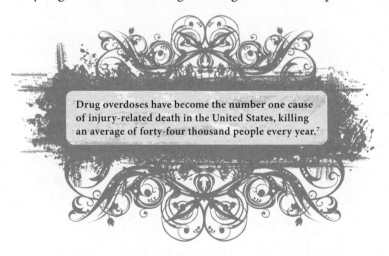

Drug overdoses have become the number one cause of injury-related death in the United States, killing an average of forty-four thousand people every year.[7]

Parallel universes are a theme in the Dark Tower series but are they possible? To understand the theory of a multiverse, we first need to understand a bit about the big bang theory. Scientists have studied how after this great explosion universes expanded, then cooled down and galaxies were brought together. In the 1980s, a physicist named Alan Guth introduced the idea of inflation. He suggested that gravity can sometimes act repulsively and it is because of this that the big bang was born. When we look at inflation creating the universe, we can predict what outcome it would have. Guth explains that inflation allows us to understand the uniformity of our universe. The observable universe is very uniform and as it expands, the uniformity would inevitably continue.[8]

String theory is another postulation of the multiverse possibility. Everything in the universe is made of matter and as we break it down we see molecules, atoms, and then electrons which are in a quantum cloud around a nucleus. These nuclei have protons and neutrons which then contain quarks. String theory posits that if you look inside those quarks you will find a stringlike filament of vibrating energy that can interact with all types of matter. This, in theory, adds extra dimensions of space time. String theory has been successful in explaining black holes and quark-gluon plasma[9] but has left scientists divided over the number of multiverses that could exist. There's no doubt that these theories will be studied for years to come.

"Go, then. There are other worlds than these,"[10] Jake tells the Gunslinger before falling to his death. Go, readers. There are so many worlds and stories that Stephen King has created. Explore the Dark Tower series if you haven't already.

CHAPTER TWELVE

Misery

In any good story there is the worthy, empathetic protagonist; think Roland Deschain in the Dark Tower series, or the kids who make up the Losers Club in *It*. On the flip side there is the antagonist; the malicious monster like Cujo, or the calculated ghoul of Leland Gaunt in *Needful Things*.

In *Misery*, these aspects of good and evil are muddier. Especially in the beginning of the novel, when Paul Sheldon is saved from the wreck of his car by what seems like a caring and concerned nurse. But, as constant readers have come to expect, no matter how harmless Annie Wilkes *looks*, she is pure venom; a woman who not only has a questionable hold on her sanity, but also has a history of killing her patients.

Author Paul Sheldon is in a unique situation. Because of his severe injuries, he has to rely on a stranger. Since he is the protagonist of a Stephen King novel, this stranger naturally wants to kill him. Or, as well remembered in the brutal iconic film version, at least maim him! Much like Jessie Burlingame in King's novel *Gerald's Game* (1992), Paul is tortured by his inability to control his environment. He is trapped and must rely on his wits, rather than any physical prowess, in order to escape.

In the modern age, we are used to immediate medical attention. Yet, Annie manipulates the dire situation of a car crash in a snowstorm to her benefit. As Paul Sheldon's "number one fan," she is displeased by the trajectory of his career, and decides to resurrect her favorite character, Misery. She holds complete control over him, as he desperately needs pain medication to minimize his suffering.

Since Paul is far from the care of a mentally sane medical worker, we started to wonder how broken limbs were tended to in years past.

Sally Mapp was a famous bone setter in eighteenth-century UK. Known as "Crazy Sally," she earned up to a hundred guineas a year because of her impressive skills.

Through writings, researchers have been able to discover how the rather common accident of a broken limb was tended to. In the Netherlands, an audience can watch a reenactment at the Archeon museum. Based on their understanding of the medieval times, "archeo-interpreters" as they call themselves, showcase what would happen to a man with a broken leg. By studying the thirteenth-century manual *Cyrurgia*, by surgeon Jan Yperman, they perfectly reconstructed the medieval bone-setting contraption, down to the same ash wood.

> The *Cyrurgia* divided the traumatic conditions and treatments of the human body into seven chapters, ranging from the head to the feet. In the seventh and final chapter "from the neck and throat down," Yperman addressed the fracture treatment. If necessary, the texts from *Cyrurgia* were supplemented with information from the book about the surgery of Yperman's contemporary colleague, the Parisian surgeon Guy de Chauliac, who lived from 1300 till 1368.[1]

According to the text, two people hold the rectangular device on either end of the broken limb, while the surgeon massages the offending bone back into place. Interestingly, after twelve days of healing, a cast was applied, even as far back as 1350! This cast was made with linen scraps covered in wax, lard, and white resin heated in a pan.

These acts of bone setting, where the bone would be painfully set back with force, continued for centuries. By the sixteenth century, bone setters were sent to work with soldiers. Often, if surgeons with these abilities were not available, blacksmiths were relied upon for the brutal wrenching of bones. As the nineteenth century dawned, there were more regulations about who could properly help those with broken limbs. Because of the 1858 Medical Act, surgeons had to be schooled and registered, and so blacksmiths and midwives were no longer able to set bones for a fee.

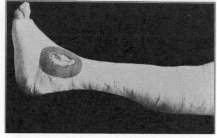

In 2005, an excavation in Colorado revealed evidence of hobbling in remains from the year 800.

Annie Wilkes, of course, was not so concerned with any regulations, as she finds every excuse not to provide Paul Sheldon with proper care. In fact, pain relief becomes the ultimate bribery. As described, we have come a long way from the archaic methods to the opioids used in *Misery*:

> Bone setting could be extremely painful, and pain was excruciating during amputations. Before 1853, only a few substances were available to dull pain, but these efforts were generally unsuccessful and many surgeons relied on their patients to faint from pain as a method of relief. A person in shock would feel less pain and bleed less, for their lower blood pressure would reduce the flow of blood, in the case of a jagged bone. Methods of pain control included: icing the limb, prescribing laudanum, drinking alcohol, and providing nerve compression or hypnosis. Icing the limb was problematic in that carting ice was a hugely expensive and laborious procedure, and storage through the warm months required ice houses and was available to only a few.[2]

Unfortunately for Paul, Annie is not as helpful as even the less than scientific caretakers of centuries past. She considers herself his biggest fan, yet her violent actions reflect disdain. This phenomenon has been proven to exist, as many celebrities have fallen prey to their self-professed fans. In 1995, singer Selena was murdered at age twenty-three at the hands of her fan club president, Yolanda Saldivar. Star of the TV series *My Sister Sam* (1986–1988), Rebecca Schaeffer, was shot to death on her doorstep in 1989 by Robert Bardo who had written her numerous fan letters. There are many examples of obsessive fan culture bleeding into violence. Rob Reiner, the director of the 1990 film adaptation *Misery*, explained the thin line between ardent follower and vicious murderer. "You definitely see in this film why fan is short for fanatic. It's tricky, because to some degree, getting attention is a real compliment. But if you go one step farther . . . When you're an artist, you need an audience. You want people to love you. You just don't want them to love you too much."[3]

In 1990, one year after Rebecca Schaeffer's tragic murder, the first anti-stalking law passed in California. It is now recognized in all fifty states.

So, what is the cause of nurse Annie's mental breakdown from devoted fan to torturer? There are several clues in the novel to what can be diagnosed as schizophrenia. This mental ailment has long been confused with dissociative identity disorder, but it is important to note that a person with schizophrenia is not full of disparate personalities. In fact, there are many false ideas of what schizophrenia is and how it affects those diagnosed. In an article for *Psychiatry Today*, researchers came to the conclusion that "Schizophrenia is the most common illness used today as a metaphor in the media and routinely appears associated with crime and violence with no medical or scientific rigor, reinforcing the stigma against this disorder."[4] They pointed to the use of the word "schizophrenic" as used in metaphor to describe something crazy, absurd, or hard to understand. In fact, data suggests that when people hear that a violent attack by someone with schizophrenia has occurred, they distance themselves from others with the disorder.

Matthias Angermeyer and H. Matschinger were among the first to report a population study of the direct effects of media reports of violent incidents on public stigma in Germany. In this study, the reported incident involved a patient with schizophrenia who attacked prominent German politicians. A comparison of pre-incident and post-incident survey responses suggested a significant increase in the social distance of the public toward people with schizophrenia, as well as an increase in the public belief that patients with psychiatric conditions were "dangerous" and "unpredictable."[5]

Socialite and dancer Zelda Fitzgerald, wife of famous author F. Scott Fitzgerald, was diagnosed with schizophrenia at age thirty.

So, what is schizophrenia, really? It is a mental illness characterized by bouts of psychosis. This psychosis manifests differently for everyone, though often comes in the form of hallucination, whether auditory or visual, as well as in the form of disorganized thoughts. With therapy and medication, those living with schizophrenia can improve greatly. Over the decades, doctors have come to understand it as more of an umbrella disease which other conditions fall under, like Schizoaffective Disorder, which is another likely candidate for Annie Wilkes. Why? Because those suffering from this disorder are known to have unstable moods. We're sure Paul Sheldon would attest to that symptom, especially when she slices off both his foot and thumb!

Unfortunately, the stigma of mental illnesses often negatively affects those suffering. The truth is, it is debatable if schizophrenia and violence go hand in hand. It has been associated with higher incidents, yet many of those involve the use of drugs. Researchers urge us to be cautious when equating mental disorders to murder. "Public stigma is among the most salient barriers to the recovery of people with psychosis and often has negative psychological effects on both patients and their caregivers." [6]

While the cause of Annie's psychotic violence is never specifically expressed, the most terrifying element in *Misery* is the reality of Annie's complex and intense bond to Paul. Unlike paranormal monsters like Pennywise, she is all human, a mirror of what can become of a warped fan's love gone too far. This is especially authentic for the author himself, as Stephen King was inspired to write *Misery* after many fans hankering for horror were less than pleased when King's fantasy *The Eyes of the Dragon* was published in 1984. Stephen King, thankfully, has not been trapped in the clutches of Annie Wilkes.

Paul Sheldon wishes he were so lucky.

CHAPTER THIRTEEN
The Tommyknockers

For an author as prolific as Stephen King, there naturally comes novels or stories that are not as well received as others. But in the case of *The Tommyknockers*, it is the author himself who is tough on the 1987 science-fiction novel. At the time that he created *The Tommyknockers*, King was suffering from a severe addiction to cocaine. Dependence on cocaine can cause extreme mood swings, paranoia, and moments of high energy. In 2014, Andy Greene of *Rolling Stone* asked King about that era in his life, asking "Did the quality of your writing start to go down?" King answered:

> Yeah, it did. I mean, *The Tommyknockers* is an awful book. That was the last one I wrote before I cleaned up my act. And I've thought about it a lot lately and said to myself, 'There's really a good book in here, underneath all the sort of spurious energy that cocaine provides, and I ought to go back.' The book is about seven hundred pages long, and I'm thinking, 'there's probably a good three-hundred-and-fifty-page novel in there.'[1]

Despite King's harsh criticism of *The Tommyknockers*, many constant readers have found appeal in one of the author's few sci-fi works. NBC produced a miniseries in 1993 starring Marg Helgenberger and Jimmy Smits. In 2018, writer and director James Wan (*Saw* [2004], *The Conjuring* [2013]) announced he was working on a film adaptation of *The Tommyknockers*.

Like most works of creativity, there were influences that helped carve the narrative of *The Tommyknockers*. Stephen King pointed to the 1927 H.P. Lovecraft story "The Colour Out of Space." In the story, a meteorite is found that has caused every living thing in a local area to become deformed, or die. This meteorite stumps scientists, and eventually it is agreed that it must be alien in origin. This mirrors *The Tommyknockers*, in which Bobbi Anderson stumbles across a mysterious piece of metal in the dirt behind her home. As Bobbi digs up what we learn is a piece of an alien ship, the denizens of Haven, Maine, become exceedingly crazy, influenced much like the people in "The Colour Out of Space." When he wrote the story, which became one of his most famous, H.P. Lovecraft pointed to his own inspirations. He mentioned the building of the Scituate Reservoir in Rhode Island, in which the town of Scituate was flooded, causing buildings to erode and a quarter of the population to move away.

One recent news development in the 1920s that was connected to "The Colour Out of Space" was the scathing truth about the Radium Girls written in the *New York Times*. The Radium Girls were a group of women, mostly young and of the lower-middle class, who were employed by the US Radium Corporation to paint clock dials with luminous paint. This paint was absolutely brimming with deadly radium, a substance that the women didn't know was going to cause them to die painful, gruesome deaths. In a review for the book *Radium Girls* (2017) by Kate Moore, the ignorance of the effects of radium is on full display:

> The workplace was considered a studio rather than a factory and provided better-than-average wages, as well as accommodated the luxury of openly conversing with other workers as long as each individual's daily production did not suffer. As each woman's level of proficiency and productivity increased, she became more valuable to the company, which resulted in wage increases; however, as the women soon learned, their achievements came with unbelievable consequences. Their employment brought various rewards as the girls and women delighted in the ability to purchase trendy clothes and, during evening socialization, received admiration from peers because their hair and clothing seemed

to sparkle in the night: a direct result of the radium powder that permeated the air within the studios. Some of the young ladies delighted in the whimsical novelty of painting their fingernails or even their teeth with the paint.[2]

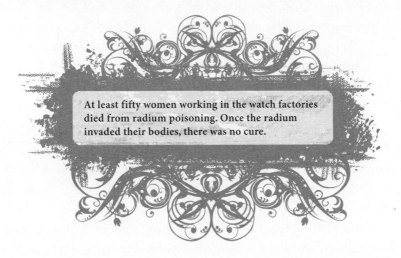

At least fifty women working in the watch factories died from radium poisoning. Once the radium invaded their bodies, there was no cure.

The true horror was that the corporation came to have the knowledge that these young women were dying, yet they did not reveal the truth, allowing more women to suffer. The radium that they had been breathing in, even rubbing on their hair and teeth, made them develop anemia, holes in their bones, and necrosis of their jaw and other body parts. This necrosis deformed their body, making them unable to walk, work, or speak.

The poisoning in the air thanks to the alien spacecraft in *The Tommyknockers* brings a similar vibe. It comes from a particular object, has no smell or clue that it is harming those nearby, yet has devastating effects. There is, of course, a paranormal element of telepathy, which is shared between those poisoned. This led us to question whether objects have been found on Earth that have been deemed alien or unidentifiable?

We found that there are objects known as extraterrestrial materials. These refer to objects derived from space that are now on Earth, including meteorites, as well as lunar samples taken by astronauts and brought home. Even particles, like the dust collected in 2019 in Antarctica from

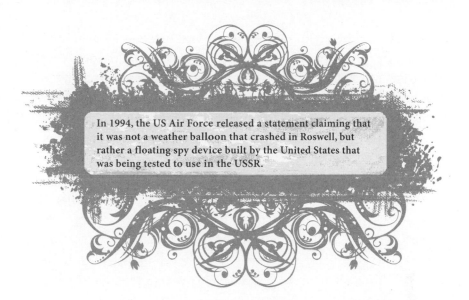

In 1994, the US Air Force released a statement claiming that it was not a weather balloon that crashed in Roswell, but rather a floating spy device built by the United States that was being tested to use in the USSR.

the Local Interstellar Cloud (a cloud which our solar system is moving through) is considered extraterrestrial material.

But what about something a little more exciting? Like, say, proof of an alien craft or technology on Earth? When most of us think of aliens on Earth, aside from the Hollywood treatment, we may think back on the incident at Roswell, New Mexico, in 1947. Today, Roswell's identity is steeped in aliens, as it has become a tourist hot spot for those fascinated with little green men. So, what exactly happened on a summer evening in rural New Mexico in 1947 that has spawned numerous movies, books, and conspiracy theories? On the town's website they offer a brief history:

In 1952, a CIA group called the Psychological Strategy Board concluded that, when it came to UFOs, the American public was dangerously gullible and prone to "hysterical mass behavior."[3]

The debris recovered by rancher WW Mack Brazel was gathered by the military from the Roswell Army Air Field under the direction of base intelligence officer Major Jesse Marcel. On July 8, 1947, public information officer Lt. Walter Haut issued a press release under orders from base commander Col. William Blanchard, which said basically that we have in our possession a flying saucer. The next day another press release was issued, this time from Gen. Roger Ramey, stating it was a weather balloon. That was the start of the best known and well-documented UFO cover-up. Once it became public, the event known as The Roswell Incident—the crash of an alleged flying saucer, the recovery of debris and bodies and the ensuing cover-up by the military—was of such magnitude and so shrouded in mystery that, seventy years later, there are still more questions than answers.[4]

There are a number of theories of what really fell from the sky. Some point to "dummy drops" in which the Air Force conducted experiments to "test ways for pilots to survive falls from high altitudes, sent bandaged, featureless dummies with latex 'skin' and aluminum 'bones'—dummies that looked an awful lot like space aliens were supposed to—falling from the sky onto the ground."[5]

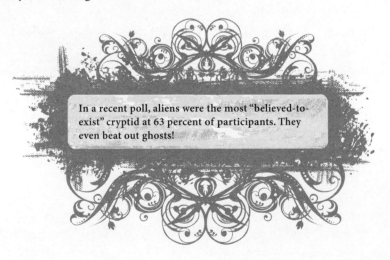

In a recent poll, aliens were the most "believed-to-exist" cryptid at 63 percent of participants. They even beat out ghosts!

While military experiments or weather balloons are more likely, it's a lot more fun to imagine that interlopers from another planet found themselves in the expanse of New Mexico. There is no concrete, scientific proof of what crashed in the desert, only witness testimony and speculation. Because of the government's evasiveness, the conspiracies grew, and thus Roswell, epicenter of alien curiosity, came to be.

In *The Tommyknockers*, Bobbi is not so lucky when she finds the crashed ship. Its alien power overcomes her, taking away her true self, leaving her a husk who only works to serve the aliens. She is joined by many others, transfixed, even hypnotized by the sentient power. The fascination with aliens is not so different. The notion of sharing our universe with others, different yet similar, captivates us. Though they may bring death and destruction, like in *The Tommyknockers*, that doesn't stop us from pursuing aliens. Whether they are speaking to us through radio transmissions, visiting us at night, or crash landing in our backyards, we want to believe.

CHAPTER FOURTEEN

The Dark Half

After the release of *The Tommyknockers* in 1987, a profound change altered both Stephen King's career and personal life. Faced with the destructive nature of his addictions, King was met with an intervention orchestrated by his wife, Tabitha. For two years, King focused on letting go of his dependence on cocaine and uniting his family. What came next was a personal project, about a man not unlike King, who was driven to madness by his two selves. He must destroy his own evil tendencies, characterized by his pseudonym come to life, in order to save all who he loves. Published in 1989, *The Dark Half* was described by James Smythe for his series "Re-Reading Stephen King" for the *Guardian* as "a novel that manages to encapsulate all King's demons—his addictions, his worries about his family life, the ups and downs of his own publishing career—while being unlike anything he'd written before."[1]

Centering on author Thad Beaumont, a father and former alcoholic, *The Dark Half* is an exploration of our disparate selves. As Smythe

One cause of auditory hallucination is Musical Ear Syndrome (MES) in which people with hearing loss purport to hear music, though there is no external source.

points out, it would be difficult not to draw comparisons between Stephen King and his character. Thad writes esoteric literature, worthy of critical praise, yet his novels receive little interest and disappointing sales. Therefore, Thad begins to write about the tales of Alexis Machine, a rough-and-tumble gangster who is the macho male ideal, right down to the bumper sticker on his slick, black Toronado which reads HIGH-TONED SON-OF-A-BITCH.

These highly successful novels about Alexis Machine's violent exploits are written under a pseudonym, George Stark, which allows Thad to separate his two distinct writing styles. Despite the money that George Stark's pulp novels provide, Thad has come to loathe this aspect of himself. Alexis Machine's casual cruelty has taken a toll, and he believes it is time to give up Stark and focus on his literary novels. Thad's wife, Elizabeth, could not agree more, as she is disturbed by Thad's behavior when he works on the Machine novels. Sometimes, it feels to her as though he is inhabited by another soul, someone with a quick temper and a sullen attitude.

There is another impetus to the metaphorical death of George Stark, and that is the inclusion of character Fredrick Clawson. Frederick has read novels by both Thad and George, and has discovered an undeniable connection in tone and style. He has uncovered that George Stark and Thad Beaumont are one and the same, and wants to cash in on this new-found information with a spot of blackmail. Rather than allow Clawson to expose him, Thad takes the matter into his own hands, inviting *People* magazine to reveal the truth with a splashy photo shoot of Thad and Elizabeth "burying" George Stark in a mock funeral.

While Clawson, later killed in a brutal and bloody manner by the physical manifestation of George Stark, is fictional, he is based on a real person in King's life. Several years before the release of *The Dark Half*, a book clerk named Steve Brown made the exact same sort of connection as Clawson. After reading the Richard Bachman novel *Thinner* in 1984, Brown came to draw comparisons between Bachman and King that led him on a trail of research all the way to the Library of Congress to search through copyright documents. With proof in hand that Stephen King was indeed Richard Bachman, Brown sent a letter to the novelist, asking if he could write an article on his literary discovery. Steve Brown described what happened next in his article "Bachman Exposed":

> Two weeks went by. Then I heard a page over the intercom at the big bookstore I worked in. "Steve Brown. Call for Steve Brown on line five." I picked it up and a voice said, "Steve Brown? This is Steve King. All right. You know I'm Bachman. I know I'm Bachman. What are we going to do about it? Let's talk." We chatted for a while and he gave me his unlisted home phone and told me to call him in the evening. I ran out and got a tape recorder with a telephone attachment and interviewed him for three nights straight over the phone. He was very relaxed and very funny throughout. He didn't seem at all upset that I had found him out. He was extremely gracious and said that he wouldn't talk to anyone else but me (outside of simply admitting it), that mine would be the only lengthy interview on the subject. It took a while for me to get it in shape and find a publisher. During this time King kept in contact and told me that more and more people had read *Thinner* and were coming after him. Finally, I published it in the *Washington Post*. From there, it went everywhere.[2]

While Steve Brown is quick to point out that he never "cashed in" on the reveal, Frederick Clawson is not so moral. He pays dearly for his blackmailing plot when George Stark, an amalgam of Thad and Alexis Machine, visits with a straight razor.

The "birth" of George Stark, a physical rendering of a concept who comes to life to seek violent revenge, occurs in the dirt of the Castle Rock cemetery. Or so it seems. At the beginning of *The Dark Half*, readers are clued into a significant event in Thad's childhood. As a ten-year-old, Thad began to suffer from debilitating headaches which grew worse. He even described auditory hallucinations, specifically of birds flapping, to his terrified mother. Diagnosed with a benign brain tumor, which oddly grew bigger as he learned the craft of writing, Thad undergoes a surgery to relieve the headaches. What the doctors find in little Thad's skull is so startling that a seasoned nurse drops a scalpel and shrieks. It becomes clear that in utero, Thad has absorbed his twin, a boy who will never live. The poor creature's eyeball, somehow sentient, stares out at the surgeon from Thad's brain before most of it is removed in haste.

Thad's parents neglect to tell him about the true nature of his tumor, which he comes to understand later is the catalyst for the birth of George Stark. This naturally led us down the path of the scientific phenomenon of parasitic twins, prompting us to separate fact from fiction.

Born in 1811, in what is considered modern-day Thailand, twin brothers Chang and Eng Bunker became "two of the nineteenth century's most studied human beings."[3] Conjoined at the chest, the brothers were immediately considered ripe for the freak show market. Discovered in Siam by a British explorer, the young men were soon traveling the globe to the delight and fascination of audiences. The term "Siamese Twins" was then used for all twins born with fused body parts, until the more accurate term, conjoined twins, came about in the lexicon in the latter end of the twentieth century. To fully understand how unique conjoined twins are, one must study the numbers. Occurring only

Chang and Eng Bunker pictured in 1874.

Chang and Eng Bunker both died on January 17th, 1874, at the age of sixty-two. Chang died first, leaving Eng to linger for two hours, knowing his death would soon follow.

once between 49,000 to 189,000 births, about half of those are stillborn, while another 30 percent die within twenty-four hours.

Perhaps because of their inherent rarity, conjoined twins have garnered much attention, even appearing on reality television series *Abby and Brittany* (2012) in which Abby and Brittany Hensel give insight into their lives as Dicephalic Parapagus twins. This term is used to describe their conjoined nature, as sharing a torso and each inhabiting one head, one arm, and one leg. It is extremely rare for twins of this particular diagnosis to live past childbirth, Abby and Brittany being two of the lucky few.

While Chang and Eng lived full lives, as do Abby and Brittany as college graduates and school teachers, the focus of *The Dark Half* is on parasitic twins, when one of the conjoined twins is absorbed by their heartier partner in the womb, leaving behind body parts and nothing more. Of course, in the fictionalized account, this leads to murder and mayhem, in the true Stephen King fashion we've come to love.

So, what causes a parasitic twin to occur? According to Wikipedia:

> Parasitic twins occur when a twin embryo begins developing in utero, but the pair does not fully separate, and one embryo maintains dominant development at the expense of its twin. Unlike conjoined twins, one ceases development during gestation and is vestigial to a mostly fully formed, otherwise healthy individual twin. The undeveloped twin is defined as parasitic, rather than conjoined, because it is incompletely formed or wholly dependent on the body functions of the complete fetus. The independent twin is called the autosite.

In *The Dark Half,* Thad is the autosite, and the remains of the twin in his brain, who later takes on the identity of his pseudonym George Stark, is the parasite. The 1989 novel, and its 1993 film version starring Timothy Hutton, are hardly the first or only appearances of such a phenomenon in media. The low-budget horror film *Basket Case* (1982) centers on Duane (Kevin VanHentenryk), who carries around his removed parasitic twin in a basket as he seeks his fortune in New York City. As you might have guessed, Duane's deformed twin also has a hankering for mischief and murder! This same theme is also found in "Humbug," a 1995 episode of *The X-Files* (1993–2018), when once again a parasitic twin, thought to have no autonomy, detaches from his autosite every night to feast on the unsuspecting denizens of a trailer park.

The reality is that babies born with parasitic twins in the modern era no longer have to endure a demeaning life in the freak show circuit. Because of knowledgeable surgeons and advances in both technology and societal understanding, children can have their extra appendages removed or altered so that they can live full, happy lives. But, while this is wholly possible, the surgeries needed to detach the autosite from their parasitic twin can be life-threatening. In the case of ten-month-old Dominique in 2017, her parasitic twin was not just an anomaly of physical appearance, it was risking her life by demanding too much blood from her small heart. Born with her twin's waist, legs, and feet growing from her back, there was a strain on Dominique's spine as well. From Côte d'Ivoire, Dominique was fortunate to receive help from a host family in the United States, where she was given a grueling surgery at the Advocate Children's Hospital in Illinois. This complex procedure included consultation from over fifty physicians before it even began. As described to Ashley Strickland for CNN, this was not a rote task for the highly accomplished staff.

> "It allowed us to come up with a plan of attack how we could safely and effectively remove this very complex part attached to this little baby's spine and end up with a healthy and happy child at the end of the day," said Dr. Frank Vicari, a pediatric plastic and reconstructive surgeon at Advocate. The team staged a mock operation to figure out who would be doing what at specific parts of the procedure. On March 8, the team worked for six hours to remove the entirety of the parasitic twin. They had to be careful to disconnect any nerves and blood vessels so that Dominique wouldn't sustain damage, numbness, or paralysis. Through extensive planning, they were even able to remove it all in one piece. She is now two pounds lighter.[4]

After a successful surgery, Dominique was able to return home to her family in Cote d'Ivoire. Her doctors believe she will no longer be hindered by her unusual condition.

Another true case of parasitic twins recently occurred in India. Similar to the fictional case of Thad Beaumont, an unnamed teenage girl was born and grew through childhood without any clue as to the interloper in her body. Like Thad, the teenager, who complained of a heavy abdomen and feeling "full," was actually suffering from "fetus in fetu." This is when the parasitic tissue is found inside the body and does not present externally. Journalist Diane Galistan explained the rare surgery for the *International Business Times*. "The lump on her abdomen had been steadily expanding over the last five years. When surgically removed, the mass was approximately two-thirds the size of a full-term baby. And was composed of hairy, cheesy material, multiple teeth, and structures resembling limb buds."[5]

Scientists have varied opinions on whether fetus in fetu is the absorption of a twin in the embryonic stage or simply a teratoma, an advanced tumor consisting of hair, skin, and bone. Obviously, Stephen King fancied the former explanation in his weaving of the horrific story of Thad and George.

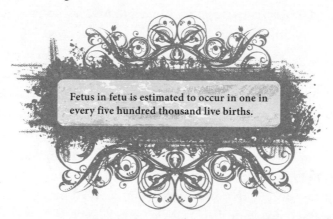

Fetus in fetu is estimated to occur in one in every five hundred thousand live births.

While science is abundant in *The Dark Half*, there is also fascinating folklore. This includes that of the psychopomp. Represented by sparrows in the novel, a psychopomp is a symbol or deity of crossing over, usually from life to death. Varying in form in different cultures, the psychopomp may be an angel, like in Islam, or come in the form of deceased family members, as in Filipino culture, where they are believed to stand at the end of the bed of the dying, waiting to usher them to the afterlife.

In Emily Dickinson's iconic poem "I Heard a Fly Buzz When I Died "(1896) a common housefly stands in as the psychopomp as the narrator struggles in the last moments of their life;

> "There interposed a Fly—
> With Blue—uncertain—stumbling Buzz—
> Between the light—and me—
> And then the Windows failed—and then
> I could not see to see—"

Psychopomps are not always the connection between life and death. In Jungian philosophy, they are represented by a wise man, woman, or even animal, and they are there to assist you between consciousness and unconsciousness. This archetype could appear in dreams, or perhaps as a spiritual guide as you seek harmony in your mind. It is also important to note that humans can choose to be psychopomps themselves, as explained at psychopomps.org:

> There is also a growing number of people who are once again learning how to fulfill the sacred role of the psychopomp. Some choose to offer their assistance in conjunction with their function as a hospice worker, or as a midwife to the dying. Others prefer to focus more on helping those who may be trapped in the spirit realms, and go by such titles as soul rescuer, death walker, spiritual guide, or shaman. There are also individuals who quietly offer aid to those in transition as they go about their routine jobs in hospitals, nursing homes, and other such locations.

The most well-known cultural icon of the psychopomp is the Grim Reaper, a nonjudgmental bringer of death who ferries the dying to wherever he or she is meant to go. Although the foreboding reaper with scythe in hand brings fear to whomever may see him, swarms of seemingly harmless birds, whether crows, sparrows, or others, are often seen as a similar bridge between our two worlds. The significance of the sparrows is explained to Thad by fellow professor Rawlie DeLesseps in his connection to George Stark, his inexplicably living twin. The sparrows have returned because Stark has come to the world where he does not belong, and must be brought back. Remember, Thad begins to hear the sparrows when his creative writing takes hold in youth, and again the insistent sound of wings and caws come when he tries to symbolically "kill" his alter-ego, signifying times of transition.

The reader finds out later in the book that when Thad goes under the knife as a child to remove the pesky tumor that has been causing his headaches and auditory hallucinations of sparrows, a swarm of real birds attacks the hospital as though directed by Alfred Hitchcock himself. Not only does the sound return to Thad as the novel continues and the bond between him and his dead twin strengthens, but at the climax of the novel sparrows cover the entirety of the Beaumont vacation home, killing George Stark and injuring the others with their sharp beaks and talons. The sparrows do their work, ushering the twin who was never meant to live back to the afterlife one painful peck at a time. Imagine if they carried small scythes like their Grim Reaper counterpart—even Stark would have to admit that would make a cute (though deadly) picture!

SECTION THREE
The 1990s

CHAPTER FIFTEEN

Needful Things

Because of the vivid world Stephen King has created for his constant readers, there is much to know about Castle Rock. One of his three most famous fictional towns in Maine, along with Jerusalem's Lot, and Derry, Castle Rock is a charming town of less than two thousand residents. It is about forty minutes from the state capitol, and is surrounded by the lush forests and lakes that vacationers have come to expect in Maine.

In the business district of Castle Rock, you may have a slice of pie at Nan's Lunchette, or perhaps stop in for yarn at You Sew and Sew. However, as we've come to love King's Maine, there is more than meets the eye. Castle Rock is the home to a murderous dog named Cujo, as well as the foreboding edifice of Shawshank Prison. So, when an antique shop named Needful Things announces its grand opening, both the residents, as well as the readers, know there is something special unfolding.

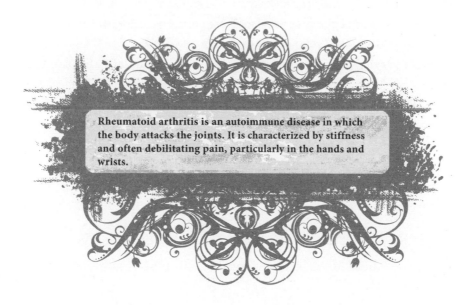

Rheumatoid arthritis is an autoimmune disease in which the body attacks the joints. It is characterized by stiffness and often debilitating pain, particularly in the hands and wrists.

Shop proprietor Leland Gaunt is considered one of King's most terrifying villains. Although at first glance he may look like a harmless old man, many people in Castle Rock are unnerved by both his stolid glance and the feel of his bony hand on their skin. As the novel continues, Gaunt's evil nature is exposed, and by the climax, in which Sheriff Alan Pangborn must vanquish him from Castle Rock, it is clear that Leland Gaunt is either a demon, or the Devil himself.

Hungry for souls, Leland Gaunt has discovered a way to convince the unsuspecting residents to give them up. He simply offers them an item, a manifestation of what they desire. For Polly Chambers, it is a necklace that inexplicably heals her crippling rheumatoid arthritis. For eleven-year-old Brian Rusk, it is a priceless Sandy Koufax baseball card that will make his collection the best in town. One fascinating find is a supposed piece of Noah's Ark, that when touched, makes the owner feel as if she is really there, perceiving the swaying of the ship and the murmur of the animals. Religious school teacher Sally Radcliffe easily gives up her soul, though she doesn't consciously understand the trade, in order to get her hands on this splinter of religious history.

Gaunt's presence ultimately causes chaos. While many of his customers kill themselves, others commit murder, and there are some, like Sheriff Pangborn, who are working to take back the stolen souls. This led us to examine the concept of the soul. Is it simply a religious or cultural belief, or is there real science to the soul?

A soul is defined as "a non-material essence of a living human that generates consciousness. It can also be considered as the spiritual principle of human beings; or, the moral and emotional nature of human beings."[1] In the *Journal of Religion & Psychical Research*, Dr. Donald R. Morse attempts to simplify how each religion views the soul. For instance, in Christianity, the typically held belief is that each human has one soul, and that once a person has accepted Jesus Christ as their savior, that soul will be allowed into heaven. In Islam, there is a widely held belief that the soul will stay with the body until the "end of days." At that time, there will be consequences based on whether the person was good or bad. Morse explains the more complicated concept of the soul in Hinduism:

Hinduism has a concept of a spirit-soul (Atman) and a "subtle body" (consisting of mind, intelligence, and the false ego) that upon death must be purified. To do this, the spirit-soul and the subtle body must enter another body to be reincarnated. If you led a pure life or after being cleansed through reincarnation, the subtle body is discarded and is liberated from the cycle of rebirth (moksha). With the impersonal viewpoint of Hinduism, when moksha occurs, the spirit-soul is absorbed into the all—this is bliss; then reincarnation occurs again. With the personal and predominant viewpoint, when moksha occurs, the spirit-soul with its spiritual body (a seed of which is found in your soul) goes into the spirit sky (similar to heaven) to be with God forever.[2]

Science tells us that the brain is what makes us ourselves. It is a series of neural pathways which dictates our actions and choices. And once this brain loses oxygen and blood flow from a beating heart, we cease to be. Yet, researchers who are also religious have continued to question whether there is a way to prove that we do, indeed, have a soul that lives apart from our body.

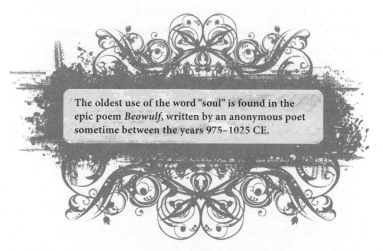

The oldest use of the word "soul" is found in the epic poem *Beowulf*, written by an anonymous poet sometime between the years 975–1025 CE.

The soul is at the heart of most religions, yet is it possible to prove such an elusive concept? In 1907, Dr. Duncan MacDougall, a Massachusetts physician, was inspired to try. He attempted to construct the most scientific of conditions possible in the era. MacDougall convinced terminally ill patients to participate. They were placed on beds that were outfitted with beam scales. The doctor would then take copious notes, including the patient's time of death, as well as make certain to include the natural fluctuations in weight that occur because of escaping oxygen and emission of sweat and urine.

At the end of his study, MacDougall concluded that there was, indeed, a weight of the human soul: twenty-one grams. This finding instantly intrigued the public, and soon the *New York Times* published an article about the experiment, calling the doctor "reputable." Just as quickly, fellow doctors and scientists argued that MacDougall's work was not sound, nor provable, as he insisted. His most egregious error was that he had much too small of a sampling, only six patients. This did not deter the doctor, who several years later, announced he would take photographs of the soul leaving the body at the time of death. He produced several images that some felt was proof of an "ethereal" light above the dying patients' skulls. However, Dr. MacDougall's work is not nearly rigorous enough to be considered proper scientific proof.

> Science writer Karl Kruszelnicki has noted that out of MacDougall's six patients only one had lost weight at the moment of death. Two of the patients were excluded from the results due to "technical difficulties," a patient lost weight but then put the weight back on and two of the other patients registered a loss of weight at death but a few minutes later lost even more weight. MacDougall did not use the six results, just the one that supported his hypothesis. According to Kruszelnicki this was a case of selective reporting as MacDougall had ignored five of the results.[3]

Despite this, MacDougall's experiment has grown into a legend that many believe proves the existence of the soul. It has even infiltrated popular culture, such as in the film *21 Grams* (2003). In an article for *Discover* magazine, Ben Thomas explains why MacDougall and those who believed in his theory at the time were driven to believe:

To understand why MacDougall wanted to weigh the soul—and why he thought he could—it helps to understand the environment in which he operated. His work is rife with terms and ideas recognizable from early psychological theorists Freud and Jung. There's a lot of talk about "psychic functions" and "animating principles"—a grasping for the precise scientific language to describe consciousness, and life itself, in a world still ignorant of FMRI [functional magnetic resonance imaging] and DNA.[1]

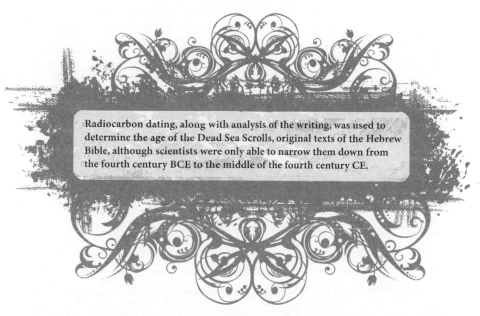

Radiocarbon dating, along with analysis of the writing, was used to determine the age of the Dead Sea Scrolls, original texts of the Hebrew Bible, although scientists were only able to narrow them down from the fourth century BCE to the middle of the fourth century CE.

While science and technology have developed leaps and bounds since the dawn of the twentieth century, humans are still curious about the intersection of religion and science. When Sally in *Needful Things* unknowingly sells her soul for what she believes is an authentic fragment of Noah's Ark, she feels as if her religious beliefs are concretely proven. The pursuit for relics is a hallowed tradition, and many archaeologists have devoted their careers to uncovering items or structures that align with religious texts. One recent example is Hezekiah's Gate:

In the Hebrew Bible, King Hezekiah despised false idols, and he destroyed everything related to his father's godless beliefs. Decades ago, a gate was found in the city of Tel Lachish, and excavations in 2016 cleared most of the structure. The six rooms covered an area of eighty by eighty foot and today stands about thirteen feet high. It doubled as a shrine and was likely also forcibly retired by Hezekiah in the eighth century B.C. According to the Bible, the gates of Tel Lachish was an important social hub where the elite would sit on benches. The building's size, as well as seats found inside, indicated that the ruins belonged to the historical shrine-gate. In an upstairs room, researchers also found two altars. Each had four horns, all of them with intentional damage, another possible sign of Hezekiah's intolerance. Elsewhere in the shrine was a stone toilet. Biblical texts mention the placement of latrines at cult centers as a method of desecration. Tests proved the toilet had never been used and was probably installed for such desecration purposes.[5]

In 2010, a group of evangelical Christians claimed to have found pieces of Noah's Ark on Mount Ararat in Turkey. Covered by snow and volcanic debris, the finding fascinated Christians across the world. If Sally Radcliffe had lived through Leland Gaunt's reign of terror in Castle Rock, she would have surely been captivated by the possibility of unearthing Noah's wooden ship. While the explorers claim to have carbon dated wood found at the site to nearly five thousand years old, many scientists are skeptical. Radiocarbon dating is a process in which the age of an object is found by measuring the radiocarbon of its organic material.

Since the Earth is proven to be 4.5 billion years old, it seems unlikely that wood carbon dated at 4,800 years old is aged enough to be Noah's Ark. Biologist and Christian creationist Todd Wood believes finding the Ark will most likely never happen. "It would have been prime timber after the flood. If you just got off the Ark, and there's no trees, what are you going to build your house out of? You've got a huge boat made of wood, so let's use that. So, I think it got torn apart and scavenged for building material basically."[6]

Whatever you believe, there is still no proof that the wood found at Mount Ararat is from the famed Ark. But, perhaps advances in archaeological and scientific practices in the twenty-first century will unearth more authentic religious relics.

While the soul cannot be proven by science in our modern era, there is no doubt that the concept itself is a hot commodity. Leland Gaunt, a demon, feeds on the souls of the innocent, much like how the vampire Barlow in *Salem's Lot* feeds

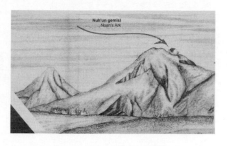

The first known image of Mount Ararat ever to appear in print. London, 1686.

on blood. While the residents of Castle Rock believed the objects they held in their hands were the most precious of items, it was really their soul, the invisible yet vital essence which makes them unique, that is the true treasure.

CHAPTER SIXTEEN
Insomnia

We've all had some sleepless nights in our lives. Whether it was forcing ourselves to be up in order to cram for a test, or staying awake with a sick child, our bodies felt it the next day. In Stephen King's *Insomnia*, Ralph Roberts is suffering from many sleepless nights and this gives him some otherworldly abilities. Spurred by a bout of insomnia himself, King spent a sleepless four months writing the book in 1990 before abandoning the project. In an interview with Wallace Stroby for *Writer's Digest* in 1991, he said, "it's a long piece of work, it's about five hundred and fifty pages long. It's no good. It's not publishable . . . the last eighty or ninety pages are wonderful. But things just don't connect, it doesn't have that novelistic roundness that it should have. And maybe someday you'll read it, but it won't be for a long time."[1] *Insomnia* did end up getting finished and was released in 1994.

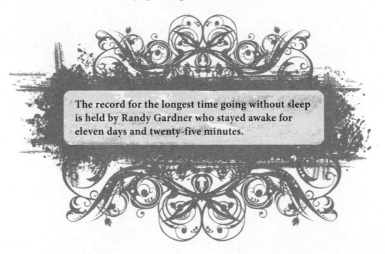

The record for the longest time going without sleep is held by Randy Gardner who stayed awake for eleven days and twenty-five minutes.

How much sleep does the human body need? Experts say adults between the ages of twenty-four and sixty-five need between seven to nine hours of sleep per night. Newborn babies and up to three months old require fourteen to seventeen hours per night while teenagers require eight to ten hours. Why is sleep so important? Our bodies use sleep as a time to repair itself and store up energy for the following day. Our minds and our moods are also affected by the quality and quantity of our sleep and sleep deprivation can have lasting negative effects.

How much sleep do other living creatures need? Cats sleep an average of fifteen hours a day while rats sleep a whopping twenty hours per day. Some animals are able to sleep using unihemispheric slow wave sleep, which allows half of their brain to stay awake while the other half sleeps. Dolphins utilize this type of sleep as well as some other aquatic mammals and birds. Imagine how much we could get done if we could utilize this strategy! Studies show that this type of sleep also improves memory, brain plasticity, and overall immune system health.[2]

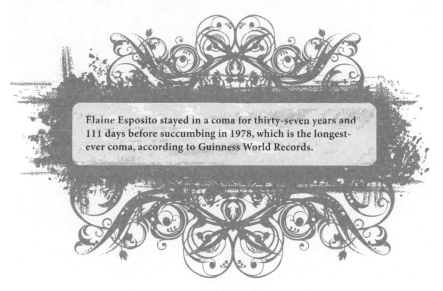

Elaine Esposito stayed in a coma for thirty-seven years and 111 days before succumbing in 1978, which is the longest-ever coma, according to Guinness World Records.

Insomnia may affect all of us from time to time but there are several different types that exist. Acute insomnia is described as a brief episode of difficulty sleeping. It could be caused by a life event that is stressful, or by travel. This type of insomnia usually resolves itself without any treatment. Chronic insomnia is a long-term pattern of difficulty sleeping. If a person has trouble falling asleep or staying asleep at least three nights per week for three months or longer, they are considered to have chronic insomnia and may need to seek help. Comorbid insomnia is insomnia that occurs with another condition. Anxiety and depression are known to be associated with changes in sleep as well as other medical conditions. Onset insomnia is a difficulty falling asleep at the beginning of the night while maintenance insomnia is the inability to stay asleep. In the book *Insomnia*, Ralph is considered to have this last form as he is unable to return to sleep.

In the novel, Ralph is able to see people's auras. Charles Webster Leadbetter is credited with popularizing the concept of auras after he studied theosophy in India in the early 1900s. He believed he could use his own clairvoyant powers to make scientific discoveries and see people's auras. (He also claimed that men come from Mars but more advanced men came from the moon!) The concept of aura photography began in 1939 by Russian scientist Semyon Kirlian. Kirlian discovered that a mysterious energy would appear when an object

Aura is defined as the distinctive atmosphere or quality that seems to surround and be generated by a person, thing, or place.

was placed on a photographic plate. The plate was connected to a source of voltage and the object seemed to be surrounded by something, which began the study of energy fields generated by living things. Kirlian published his first scientific paper on the subject in 1961, in the Russian *Journal of Scientific and Applied Photography*. We spoke with an aura photographer, Annette Bruchu, to understand more about this art.

Kelly: **"In Stephen King's book *Insomnia*, a character is able to begin seeing people's auras. Tell us how you got into aura photography and/or reading auras."**

Annette Bruchu: "'You can see aura?' is a question that I am frequently asked. The dreamer, the visionary, the one with the big imagination. Or is it my reality of seeing the space between the spaces of all the energy around everything that is alive, from people, electrical wires, animals, trees, the ground, and water flowing in the stream? As a child I was surrounded with color, imagination, laughter, and fun. This halo, what I called it, I could see around and in front of people was not only a white light, but has color. The color was stronger when the conversations were stronger or more intense, when more excitement was happening. I didn't realize that others were not seeing color aura until people would come up to me and ask me 'what color is their aura' or 'I think you need to get your eyes looked at.'"

Meg: **"That must have been eye opening, excuse the pun, for you to discover you were seeing something no one else was!"**

Annette Bruchu: "At the ages of six, seven, and eight I felt very normal and everyone else was off because they often questioned me about the halos they had. I felt like I had a pretty relaxed childhood and loved to stay up late. I would often find myself mesmerized with the TV or watching the fireplace very zoned out. As an adult, I began to watch people and their behaviors and reactions, and I began to better understand the colors by the moods people were in."

Kelly: **"How did you get into aura photography?"**

Annette Bruchu: "One day a psychic gal said to me 'why don't you buy an aura camera. You could probably better show people their colors and put it to good use.' I didn't even know there was such a camera until I began a brief study to see which camera system was the closest to what I was seeing around people. Then I bought the Kirlian digital aura camera system. I set it up and I was actually happily surprised by how accurate the technology is. My dad once said to me when I was a child that everyone has a talent and they should depend on their talent to make a living and so I did just that."

Meg: **"What is the process like if someone wants their aura read?"**

Annette Bruchu: "I am a vendor at a great many metaphysical expos around the United States where I set up the systems and people approach me to get the photo taken. They would sit comfortably in front of the computer system where they can see the camera as they place their hand on the silver sensor to make contact. The process entails sending the information of their energy through the program so that instantly they can view their aura live on the computer screen. This technique is so fascinating and it's spectacular to see how the human body shows the glow of life force's electrical fields. After a moment it all calibrates and I print several pages of information for them to take home. Also, I explain to them what it all means. It is delightful information."

Kelly: **"What do the various colors and variations mean when you are reading their aura?"**

Annette Bruchu: "Every color aura has a gift to offer and specific quality to share. The colors that show up in aura have great information. For instance, yellows show their caring side for people and nurturing with excitement and joy. I often see people drawn to them as their counselor or mentors. They are uplifting with a gift of intelligence that keeps them wanting to learn their whole life long. They have a great sense of humor and a magnetic energy that people gravitate toward to feel safe. They're great party people and planners that like a fun time. They love the limelight and work well with children and seniors."

Meg: "I know people like that! And yellow is my favorite color."

Annette Bruchu: "Greens are naturally attracted to the folks with the yellow aura. The green energies in an aura tend to be more serious personalities. They are the planners and details are important to them. They are usually in a role of science, financial, or medical, and the attraction to the yellows are strong for both of them because yellows are chatterers and greens tend to be listeners."

Kelly: "Okay, I'm definitely not a green!"

Annette Bruchu: "Violets are the sensitive observers and the visionaries. They can come up with such amazing ideas and the orange aura folks create it for them. The violets are sensitive to smells, touch, taste, and feel. And a bit emotional, sentimental, as well as observant. Oranges are creative and talented as they march to the beat of their own drum. They add a childlike innocence to our life. They are great as photographers, artists, and musicians. They are attracted to the violet auras because of the sensitive side they share. Blue is the aura of the spiritual leaders. They are the calm ones that for some reason show up, and everyone feels at ease with them. They lead people well and love the outdoors where they enjoy the air on their skin and all that mother nature offers."

Meg: **"It's fascinating to think about the people in our lives who fit so distinctly into these categories. What about the rest of the colors?"**

Annette Bruchu: "Reds are the passionate people with the drive to do it their way, so I usually say 'get out of their way.' The darker red can also show up as stress, grief, and in people who are the worriers. I call whites the supernovas. This is an interesting hue because they can carry all of the colors of the rainbow in their aura and they can be adaptable to which characteristics suit the situation. They are very talented in so many areas and they can feel depleted in their stamina very quickly. They can also be eccentric thinkers. Black auras mean there is no energy to be found. They are unplugged and unaware of their surroundings. All of the colors are magnificent to see."

Kelly: **"Can you talk more about the sort of technology and/or camera you use to read auras?"**

Annette Bruchu: "I have and use the Kirlian digital photography system. It is the one I like for a most interesting aspect; it has electro-photonic imaging where you can see live altered states of consciousness including meditation, very scared, mental stress, confusion, or laughing. It is quite diverse, and accurate."

Meg: "Do you have a favorite Stephen King movie or novel? What is it about the story that you like?"

Annette Bruchu: "I like nail biters and being frightened comes easy for me. I startle easily and I am a bit jumpy because of my senses that are sensitive. I like *The Shining* because it was a bit disturbing and it had me screaming. I am all about numbers and the room number 237 at the Stanley Hotel is a seductive number with a scene of a lovestruck woman that ended up heartbroken and ends her life in the bathtub in that room. Devastating! When you visit The Stanley, you still get yourself ready to be frightened years after the movie had been released. I am so afraid of creepy noises and weird things in old hotels, still today."

Meg: "That's one of my favorites, too!"

Annette Bruchu: "*Carrie* was one of the first Stephen King movies that I saw as a kid. I didn't sleep for weeks. I was a teenager myself and the girl had the strange powers that I could relate to. Seeing auras should be effortless when our mind and our eyes are relaxed. We open up our sensory apparatus, when color is more apparent to our physical eyes. When we get out of our own way."

Kelly: "It's incredible to me that you were going through this time in your life and were able to relate to Carrie in the movie. You both had powers!"

Our conversation with Annette Bruchu made us curious about our own auras and what would show up in our photographs!

Insomnia also explores how we view death. Ralph begins seeing "little bald doctors," two of whom want to preserve the natural order of death while a third brings chaos. How do various cultures view these harbingers of death like the Grim Reaper? In Greek mythology, the Thanatos guides souls to the next life through a nonviolent death. They believed death to be a peaceful transition into eternal rest. In Irish mythology, death is foretold through the shriek of the banshee. In the fourteenth century, it was believed that the scream of the female banshee would predict impending doom. Around this same time, the Grim Reaper became

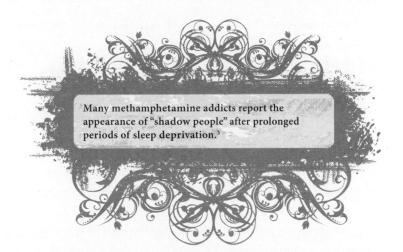

Many methamphetamine addicts report the appearance of "shadow people" after prolonged periods of sleep deprivation.[3]

popularized in society due to the catastrophic effects of the Black Plague. Death comes for Ralph at the end of *Insomnia*, although it is a sacrificial death in order to save his wife's daughter. He goes peacefully with the two bald doctors by his side.

CHAPTER SEVENTEEN

The Green Mile

In 1996, Stephen King released eight books; six of them being serialized installments of *The Green Mile*. In the 1800s and early to mid-1900s, serialization was a popular form of publishing. This format gave authors a wider readership and publishers saw greater profits. Charles Dickens is most often credited with beginning this fad with *The Pickwick Papers* (1836–1837) but authors throughout history have used this model. After Stephen King published *The Green Mile* this way, several other authors jumped on the trend including John Grisham.

The story in this novel focuses on death row, known as the "green mile," due to its linoleum floor. Paul Edgecomb is one of the guards at the prison and we follow the story through his eyes. Stephen King, on writing this story, said "the human spirit is alive and well even under the most difficult circumstances . . ."[1] and it is proven in this setting. We meet and empathize with guards, convicted killers, and an innocent inmate whose name is John Coffey.

Paul Edgecomb is touched by a bit of the magic from John Coffey and will inevitably have a long life because of it. What is the average age of life expectancy? In the United States, the average life expectancy is eighty years while in Japan it's eighty-five years. The oldest verifiable person on record was a French woman, Jeanne Calment, who lived to be 122 years old. Paul Edgecomb is 104 in the novel and he's not sure when he's going to die. Could we live longer as technological and medical discoveries advance? Not likely. Scientists say the answer lies in our bodies, not our health. Our bodies begin to break down, our DNA gets damaged, and our organs don't work as efficiently.

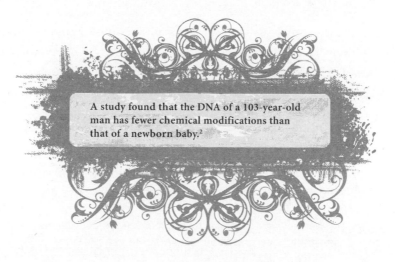

A study found that the DNA of a 103-year-old man has fewer chemical modifications than that of a newborn baby.[2]

One of the unexpected characters in the book is a mouse named Mr. Jingles. He is trained and also has magic in him after being healed by John. Is it possible to train mice? It is, but it takes patience and about two weeks of consistent practice. Mice, just like other animals, prefer to be rewarded with food for good behavior. By luring a mouse with food and getting them comfortable with you, they will be more likely to follow directions. Use a similar tone of voice when offering a command and only move on to the next trick after one is mastered. All sorts of animals can be trained to accomplish varying degrees of tricks or duties. Companion animals, show performers, and police dogs all undergo training that they learn through routine. Not all animals can be trained, though. Based on their personality or temperament, it may be easier to forego the tricks and let them be themselves!

John Coffey can be seen as a healer in *The Green Mile*. He fixes Paul's bladder infection, brings Mr. Jingles back to life, and even removes a brain tumor from a woman. What is the history behind folk healers? People over the centuries have been seen as having otherworldly powers and the ability to heal. Some women who aided in midwifery or used herbal remedies were accused of witchcraft. Others simply used the knowledge passed down to them through generations to provide health care that may not have been available otherwise. There are those who claim to be healers but are proven to be frauds. For example, psychic surgery began to gain popularity in the mid-1900s. Surgeons would supposedly be able to penetrate a person's body without surgical tools and remove any lesions or foreign objects that were making the patient sick. Investigations have found that these pseudo surgeons were actually using fake blood and animal parts with a sleight of hand to make money off desperate people.

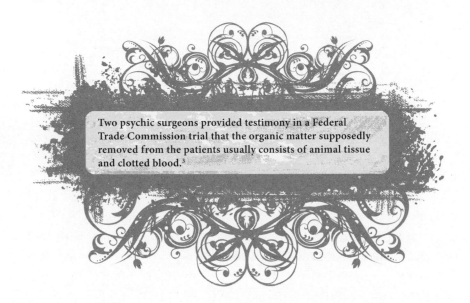

Two psychic surgeons provided testimony in a Federal Trade Commission trial that the organic matter supposedly removed from the patients usually consists of animal tissue and clotted blood.[3]

The focus of the novel is the long walk to death. In *The Green Mile*, death row means you will die by the electric chair which they nicknamed "old sparky." Capital punishment has existed for centuries and the methods for execution have ranged from quick and painless to downright barbaric. In South and Southeast Asia, it was popular to torture and execute people by crushing them with elephants. Elephants are highly intelligent and could be trained in various methods of either prolonging torture or by killing someone swiftly. This practice was used particularly in India and continued until the later part of the nineteenth century.

A practice known as the blood-eagle killing-ritual was recorded in Norse literature and was a horrific method of publicly killing someone. The method involved carving into the back of the person, removing their ribs, and taking out their lungs through the back to appear as wings. Some historians argue that the ritual is fictional and meant to induce horror in its readers. Others hypothesize that warriors left face down after battle could have fallen victim to ravenous birds and appeared to be sacrificed. Crucifixion was a method of public execution that dates back to 300 to 400 BCE. Those being crucified were often nailed or tied to the upright cross structure. Depending on the crime, the method of attaching the criminals to the cross varied and could kill them within a day or leave them to suffer for several days. Ling chi was a method of execution that is often called "the death of a thousand cuts." This was practiced in China from 900 CE until its abolishment in 1905 and was most often reserved for those who committed the most heinous crimes.

The guillotine was invented in 1789 in France as a more humane method of execution. The man who invented it, Dr. Joseph-Ignace Guillotin, was actually opposed to capital punishment but argued that this invention would be quicker and less likely to fail than the crude axe beheadings of the time. Public beheadings were a popular spectator event and continued in France until 1939. Drawing and quartering was another gruesome public execution. Those who were sentenced to death were pulled behind a horse to be

The guillotine claimed its first official victim in 1792.

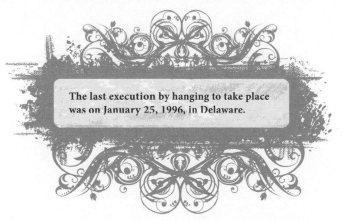

The last execution by hanging to take place was on January 25, 1996, in Delaware.

hanged, then their bodies would be pulled apart in separate directions by four horses. Firing squads began to be used as an execution method for military personnel and involved the accused standing blindfolded to be fired upon by fellow soldiers.

The electric chair is the means of execution in *The Green Mile* and its history is disturbing. The method originated in the United States and was used almost exclusively here as recently as 2013. Some states including Alabama, Florida, South Carolina, Kentucky, Tennessee, and Virginia still have the electric chair as an option for capital punishment. Invented in 1881, the electric chair was seen as a better, more humane alternative to hanging. The way the electric chair works involves several steps. First, the condemned is strapped into a wooden chair. Second, their head and legs are shaved and electrodes are attached. Finally, an electric current is sent to the body and repeated until the person dies.

The first public execution using the device did not prove successful. William Kemmler was convicted of murdering his girlfriend with a hatchet and was put to death on August 6, 1890. The first seventeen-second current of electricity caused Kemmler to become unconscious but did not kill him. A second shock caused blood vessels to rupture and bleed while the electrodes attached to him singed. The total execution took eight minutes and onlookers could hardly watch. One witness, Deputy Coroner Jenkins, said, "I would rather see ten hangings than one such execution as this. In fact, I never care to witness such a scene again. It was fearful. No humane man could witness it without the keenest agony."[4]

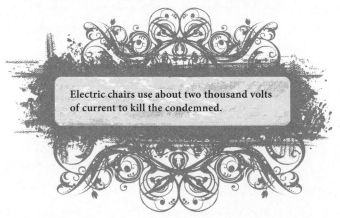

Electric chairs use about two thousand volts of current to kill the condemned.

The use of the electric chair declined over the decades as a more humane solution was sought. Lethal injection became the preferred method for execution and is still used in numerous countries. The idea was pitched in 1977 by a medical examiner named Jay Chapman. He believed three drugs should be injected in order to first cause the condemned to become unconscious, then paralyzed, then cause the heart to stop beating. This became the cocktail for lethal injections but was never medically tested or peer reviewed. Not every state uses this three-drug method but one thing is clear: opponents claim this is no less cruel or painful than previous methods of execution.

John Coffey dies by electrocution at the end of *The Green Mile* but assures Paul that his time has come. As Paul says in the novel, "we each owe a death, there are no exceptions."[5] Although he doesn't know when death will come for him, he feels at peace knowing he put down in words the story of what happened all those years ago at the Cold Mountain Penitentiary. May we all feel this sense of tranquility when our time comes.

CHAPTER EIGHTEEN

The Girl Who Loved Tom Gordon

Getting lost in the woods would be terrifying for anyone but imagine having it happen to you as a nine-year-old child. This is the premise of Stephen King's novel from 1999, *The Girl Who Loved Tom Gordon*. King compared the novel to "an unplanned pregnancy" and said the idea came to him during a Red Sox game. "Stories want only one thing: to be born. If that's inconvenient, too bad."[1] Trisha MacFarland is on a hike with her family on the Appalachian Trail and gets separated from them. She's left with only a backpack containing a Gameboy, a Walkman, a hard-boiled egg, a tuna sandwich, two Twinkies, a liter of Surge, and a poncho. She fights real menaces such as hunger, dehydration, and pneumonia but also feels like she is being stalked by the God of the Lost; a wasp-faced, evil entity. The pitcher, Tom Gordon, seems to appear to her and help guide her throughout the story.

In order to understand how to handle being lost in the woods, we spoke with *Survivorman* (2004–2015) star Les Stroud and got his advice in case we are ever in this situation.

Kelly: **"In Stephen King's book *The Girl Who Loved Tom Gordon*, Trish is lost in the woods and ends up eating some flowers that she finds. What plants and flowers are edible and safe to eat in various parts of the country?"**

Les Stroud: "There are thousands of edible wild plants across the globe and no ecosystem is void of many things you can eat. What changes is, are they in season, are you around the areas where they grow, do you know which parts are edible, do they have any poisonous look-alikes? There are hundreds of books available pointing to the plethora of plants that are edible as well."

Meg: **"The character in the novel remembers an episode of *Little House on the Prairie* (1974–1983) that recommended if lost to follow a creek. Is this good advice?"**

There are dozens of edible flowers in nature but it's always safer to check a guide before tasting.

Les Stroud: "It is hit and miss. Thousands of creeks lead to absolutely nowhere. As a general rule, sure, going downhill tends to be a good direction to take and most creeks flow into bigger creeks and into rivers and into civilization. But not all. Many people have become horribly lost because they attempted to follow a creek out to safety when in fact climbing up would've been the better choice."

Kelly: **"If someone finds themselves lost in the woods what advice would you give them?"**

Les Stroud: "First, calm down. Then, practice the Survivorman Three Areas of Assessment: Take stock of what you have close at hand (backpack with food, friends, two people injured, a tent), then take stock of what is near (fresh water, one canoe left, plenty of firewood). Then, finally assess what is further out (a cabin one mile back up river, a highway one mile due east through the bush, a trail two miles down the river's shore). Now you have a lot of information and you can make an informed decision on what to do next."

Kelly: "Meg, I'm sorry to say if you were lost in the woods with me I'd probably only have a six pack of beer and maybe some lipstick on me!"

Meg: "Just promise you'll share!"

Meg: **"The girl in the book is suffering from hunger and dehydration. What are the safety measures to take when finding water to drink in the wild?"**

Les Stroud: "Search for moving water instead of stagnant. If it is deep, try to weigh down a container and lower it to the deeper areas of the water body before pulling up quickly, as there are likely to be less pathogens there. Remember that dew in the morning can provide a lot of liquid in some places. Look for changes in vegetation in the forest as that often indicates the presence of water. Collecting rain will be safer than ground water. Travel upstream if possible so you get to the purest spot of a stream where it is likely to be less contaminated by animal feces."

Bear attacks happen about forty times per year globally.[2]

Kelly: **"The character in *The Girl Who Loved Tom Gordon* ends up in a confrontation with a bear at the end of the novel. What should people do if they encounter a bear in the woods?"**

Les Stroud: "If a black bear, play tough, make noise, try to look big, fight back . . . do all you can to scare it off. If a grizzly bear, play dead and hope it gets bored with you. In both cases, never run. If it's a polar bear . . . pray."

Meg: **"Do you have a favorite Stephen King book or movie? What is it about the story that draws you in?"**

Les Stroud: "*The Green Mile* by far. I hate that I am drawn in due to this horrible fear of being wrongly accused but I am elated when there is redemption and payback at the end."

Hopefully we will never find ourselves lost in the woods, but with Les Stroud's advice, we feel more prepared if it ever happens!

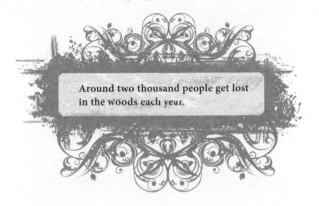

Around two thousand people get lost in the woods each year.

Trisha is suffering from hallucinations from hunger or dehydration as the days go by in the book. Does this actually happen to the human body? People can survive for a period of time without adequate food or water but our bodies begin to preserve energy and live off of fat reserves before moving on to muscle breakdown. Various bodily systems will begin to break down after just a few days without food or water and side effects may include faintness, dizziness, a drop in blood pressure, a slowing heart rate, abdominal pain, body temperature fluctuation, heart attack, or even organ failure.[3] Since children naturally have a lower fat and muscle reserve than adults, they are more at risk for developing symptoms and complications much sooner. Trisha absolutely could have been suffering the effects of starvation within this time period and hallucinating.

We've all heard the old adage that you could catch a cold if not properly dressed when outside. Trisha gets pneumonia after being lost in the woods for nine days. Is there any truth to this old wives' tale? According to Dr. Roshini Rajapaksa, you can't catch a cold from being cold but it can lead you to being more susceptible to illnesses. "Cold weather can dry out the lining of your nose, leaving you more vulnerable to an infection. Some research also suggests that prolonged exposure to the cold may suppress the immune system."[4] Trisha's pneumonia was a lung infection caused by a virus or bacteria. These germs made the air sacs in her lungs fill with fluid, phlegm, or mucous, making breathing more difficult. Pneumonia is the number one leading cause of death in the world for children under the age of five. In the United States, pneumonia tends to be less fatal for children but is the number one reason they are hospitalized.[5]

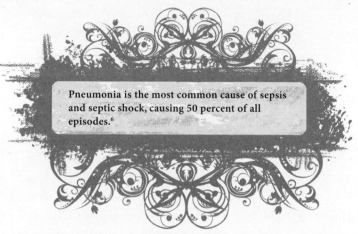

Pneumonia is the most common cause of sepsis and septic shock, causing 50 percent of all episodes.[6]

Trisha survives her encounter with the bear in the woods and is able to find some peace, maybe even faith, by the end of the book. She points up at the sky, like her hero Tom Gordon does, to signify that perhaps there is a higher power.

CHAPTER NINETEEN

Gwendy's Button Box

Throughout Stephen King's prolific career, he has written novels, short stories, and screenplays. He has even written as two men, creating his alter ego, Richard Bachman, in order to publish at an accelerated rate. He has worked with Michael Jackson on the music video "Ghosts," appeared in films like *Pet Sematary* (1989) and even directed the film adaptation of his story *Trucks* into *Maximum Overdrive* (1986). So, it should be no surprise, as a creator devoted to storytelling, that King has collaborated with fellow writers. His first cowritten novel was *The Talisman* (1984) with Peter Straub. They worked together again to bring the sequel, *Black House*, to readers in 2001. In 2017, two coauthored King works were released. *Sleeping Beauties*, written alongside King's son Owen King, is the story of a strange epidemic spreading across the globe. Focused on a

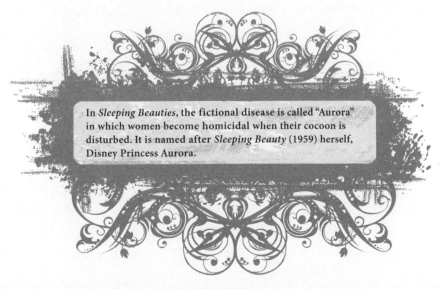

In *Sleeping Beauties*, the fictional disease is called "Aurora" in which women become homicidal when their cocoon is disturbed. It is named after *Sleeping Beauty* (1959) herself, Disney Princess Aurora.

small town in Georgia, *Sleeping Beauties* depicts a world in which women are cocooned like moths and men are left to figure out how to fix the chaos. A starkly feminist novel, it allows the women to save themselves. Owen King is currently writing the pilot script for AMC.

The other collaborative work in 2017 was the novella *Gwendy's Button Box*. Set in the 1970s in Castle Rock, *Gwendy's* was written with a longtime friend of King's, Richard Chizmar. It is the story of a middle schooler named Gwendy who is given an odd, magical box by a stranger. This box, which she keeps a secret from her parents and friends, is both a blessing and a curse. It gives her small chocolates that once eaten, make her life better. She loses her extra weight, gets straight A's, and is a star on the track team. Yet, there are buttons on this box that carry a considerable burden. The stranger who gives her the box, mysterious much like the man who gives Jack the magic beans, explains that each button corresponds with a particular continent of the world. And if pressed, chaos will ensue. A curious child, Gwendy eventually presses the button that represents South America.

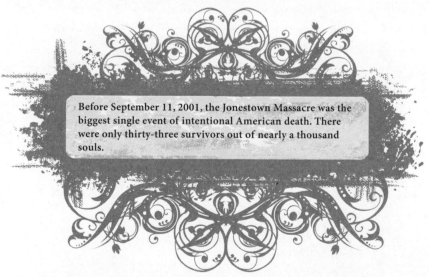

Before September 11, 2001, the Jonestown Massacre was the biggest single event of intentional American death. There were only thirty-three survivors out of nearly a thousand souls.

She finds out the next morning that the Jonestown Massacre, in which over nine hundred people perished, occurred the evening before. Convinced she is to blame, Gwendy is wracked with guilt. She can't stop thinking it is her, and not Jim Jones, who is to blame. Later, the mysterious stranger assures her it is cult leader Jim Jones, not her press of the button, that caused the deaths. This led us to question the concept of coincidence.

Many of us have experienced odd coincidences we can't explain. Some believe this is the work of God, or perhaps the cosmic wisdom of the universe. Mathematicians Persi Diaconis and Frederick Mosteller took a more scientific approach, defining coincidence as "a surprising concurrence of events, perceived as meaningfully related, with no apparent causal connection."[1] They explain in their 1989 research article "Methods For Studying Coincidences" that given the large number of humans on the planet, "with a large enough sample, any outrageous thing is likely to happen." More importantly, humans are not able to fully, mathematically appreciate what is truly unlikely. What we feel is a coincidence is rooted in our own personal belief systems. We are more interested in coincidences that have to do with us, rather than other people. In the *Atlantic* article "Coincidence and the Meaning of Life," writer Julie Beck uses the example of birthdays:

> People can be pretty liberal with what they consider coincidences. If you meet someone who shares your birthday, that seems like a fun coincidence, but you might feel the same way if you met someone who shared your mother's birthday, or your best friend's. Or if it was the day right before or after yours. So, there are several birthdays that a person could have that would feel coincidental.[2]

People who describe themselves as religious tend to believe more in coincidences.

Again, the observation of coincidences are in a way, egotistical, as Beck explains that research proves it reflects more on the person experiencing the coincidence than the actual mathematical likelihood:

Research has found that certain personality traits are linked to experiencing more coincidences—people who describe themselves as religious or spiritual, people who are self-referential (or likely to relate information from the external world back to themselves), and people who are high in meaning-seeking are all coincidence-prone. People are also likely to see coincidences when they are extremely sad, angry, or anxious.

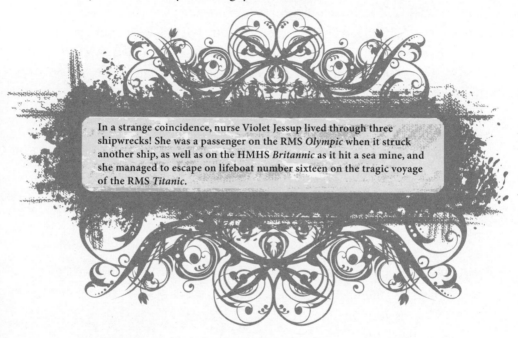

In a strange coincidence, nurse Violet Jessup lived through three shipwrecks! She was a passenger on the RMS *Olympic* when it struck another ship, as well as on the HMHS *Britannic* as it hit a sea mine, and she managed to escape on lifeboat number sixteen on the tragic voyage of the RMS *Titanic*.

In the fictional, magical world of *Gwendy's Button Box*, it's hard to say whether Gwendy pressing the button corresponding with South America really *did* set the Jonestown Massacre into motion, but as we know in the real world, it was no simple button, or coincidence, that caused such a blight on our collective history.

What is a fortunate coincidence (if we do say so ourselves!), is that we had the great honor of interviewing Richard Chizmar, coauthor of *Gwendy's Button Box*, as well as the founder of *Cemetery Dance Magazine and Publications*.

Meg: **"Writing fiction can be a very personal creative endeavor. Can you describe how *Gwendy's Button Box* came about, and how logistically you and Stephen King handled writing a book together?"**

Richard Chizmar: "Steve and I have been friends for a long time—since back in the early 1990s—and have done a lot of business in the book world, but I never dreamed I would one day write something with him. We email a lot about a variety of subjects ranging from books and movies to sports and family. One afternoon, we were emailing about round-robin books and collaborations. That general discussion led to Steve telling me about a story he had started but had been unable to finish. The next morning, *Gwendy's Button Box* showed up in my email along with a note reading, 'Do what you wish with it.' And that's how it happened."

Meg: "That's my dream!"

Richard Chizmar: "Logistically, it was a pretty simple process. I picked up where Steve left off with the story, added about ten thousand words or so, and sent it his way. He added a chunk and sent it back to me. And then we played ping-pong with the manuscript, back-and-forth, until we were finished. We each enjoyed total freedom with the direction we took the story and rewrote each other's work to find one singular voice. It was a blast."

Kelly: **"You have since written *Gwendy's Magic Feather*, solo. We imagine it would be quite an honor to "borrow" Castle Rock and infuse your own style and creativity. It's quite a unique writing challenge! How did you approach such a feat?"**

Richard Chizmar: "It was a huge honor and responsibility. On one hand, having cowritten the first book with Steve really prepared me for the challenges of *Magic Feather*. On the other hand, the second book is set in 1999, so a lot of things had changed since the end of *Button Box* (not the least of which is that Castle Rock was supposed to be completely destroyed earlier that decade at the conclusion of *Needful Things*!). My main concern was getting the facts straight—people, places, times. I knew this was sacred territory for King's constant readers, and I didn't want to mess that up. For accuracy, I relied on my friends—and longtime King experts—Bev Vincent and Brian Freeman to give the various drafts a close read. And they did a terrific job. The rest was all about letting myself get lost in the story, lost in the small Maine town of Castle Rock, and following Gwendy wherever she took me. I've come to really trust her instincts."

Meg: "**Your career has spanned short stories, screenplays, novels, and everything in between! What sort of research do you prepare when working? Is it important for you to be scientifically or historically accurate, even in supernatural projects?**"

Richard Chizmar: "It really just depends on the project. I'm writing a book now that is set back in 1988 in the town I grew up in. It's crucial to the story that I get the time period correct. Geographically, socially, economically, right down to the movies and music and trends of the time period. It's been a lot of fun and a lot of research. Other stories I've written, that kind of accuracy is not as important to the characters or plot. They could be set pretty much anywhere at any time."

Kelly: "**Since the inception of *Cemetery Dance* over thirty years ago, would you say horror literature has changed? Or, as an editor and reader, does it still come down to the same aspects of talent?**"

Richard Chizmar: "I think it still comes down to the same aspects of talent. For me, it's still all about story and narrative drive and caring about the people and places and moments that I'm reading about. Take me away to another place, another time, and (as a reader) I'm yours."

Meg: "**Do you get a particular satisfaction in discovering authors and then watching them flourish in their career? How would you describe that feeling?**"

Richard Chizmar: "I love the feeling that kind of discovery brings. How to describe what that process actually feels like? Hmmm, I'd say a mixture of pride (for that particular writer seeing a dream come true and so much hard work realized) and gratitude (that I'm in a position to help do that) and awe (that, even after all these years, it feels like I'm doing exactly what I was put on this Earth to do).

Kelly: **"Cemetery Dance Publications has published Stephen King's work, including limited editions, as well as books about him. Recently there has been talk that Stephen King's work is finding a new, younger audience, thus the sudden boom of television and film adaptations. Do you believe we are living in a 'King Renaissance' or that he has had a 'comeback' as some media assert?"**

Richard Chizmar: "I absolutely believe he is experiencing a renaissance. I often give talks at local high schools. A decade ago, the majority of the students I spoke with only knew Stephen King through the film adaptations of his work. They loved *Carrie* and *It* and *The Shining*, but knew nothing about the books they were based on. That has all changed now. The sheer number of King projects, plus online marketing and publicity, has steered it all back to the books. I love what is happening."

Meg: **"Can you tell us about your future projects? What can we look forward to?"**

Richard Chizmar: "I have a couple of graphic novels due late in 2020, as well as a relatively slim nonfiction collection and a not-so-slim collection of my first nineteen *Stephen King Revisited* essays. I'm currently working on a novel now that should (hopefully) come out sometime in 2021, as well as a collection of four novellas. Plus, as always, a handful of stories and scripts."

Kelly: **"What is your favorite work by Stephen King? The one book or story that has stayed with you, above all."**

Richard Chizmar: "My all-time favorite is *It*. Reading that novel when it first came out (I was in college at the time) cemented in my mind that this is what I wanted to do with the rest of my life. It didn't just open that door for me—it broke the door down. And I've never looked back."

We agree with Richard Chizmar that *It* is a King masterpiece! It was a pleasure to ask questions of someone who is not only a colleague of Stephen King, but also an impressive author, editor, and screenwriter in his own right.

At the end of *Gwendy's Button Box*, Gwendy, now in college, has had her time with the mysterious box and must give it up to the next person deemed suitable to handle it. She has proven herself to be noble, only using the box to murder, once, and for a justified reason. It leaves us curious to find out what happens in *Gwendy's Magic Feather*. And we can't help noticing that King has given Chizmar the same sort of responsibility, allowing him to write within the strange and enigmatic town of Castle Rock.

SECTION FOUR
The 2000s

CHAPTER TWENTY

Dreamcatcher

Stephen King was struck by a vehicle while walking in 1999 and his world changed forever. His right hip was fractured, his right leg was broken in nine places, and one of his lungs collapsed. He notes in 2001's *Dreamcatcher* that he was never so grateful to be writing than while writing this novel. Suffering from physical pain, he relished the opportunity to handwrite the entire book, saying it put him back in touch with language. The story has several supernatural elements to it including telepathy, aliens, and body possession. King recalled using oxycontin for the pain from his accident. "I was pretty stoned when I wrote it, because of the Oxy, and that's another book that shows the drugs at work."[1]

A 2017 survey reported that 47 percent of Americans believe in aliens, which is about 150 million people.[2]

The novel was turned into a movie in 2003 and although Stephen King didn't write the screenplay adaptation for it, he has adapted many of his stories for the big screen. In order to understand this process, we spoke to writer Kara Lee Corthron about her experience working on the television series *You* (2018–).

Meg: "**How do you approach adapting a novel for a television show? How does it differ in pacing compared to other writing you've done?**"

Kara Lee Corthron: "This is more a question for Sera Gamble, the series creator. I'm on a team of multiple writers and at the beginning of our work on season two, when I joined the staff, our task was to adapt Caroline Kepnes's second book, *Hidden Bodies* (2016), while holding on to all of the organic work that came out of season one. As a result, we used a lot from the book, but also changed a lot. It's tricky because once the book becomes a show, it's its own entity separate from the book. So, while respecting the original text, we have to write what serves the story best in the TV genre."

Kelly: "**Many of Stephen King's characters are dark and sinister. How do you write for a character like Joe Goldberg and balance his humanity and his monstrous nature in order for the audience to still connect with him?**"

Kara Lee Corthron: "Joe Goldberg is such a fun character to write. It's not as hard to write for him as you might think. The trick is to understand how Joe sees himself and let that be our guide. Joe sees himself as a feminist and one of the last true romantics willing to do anything for the woman he loves. All of his behavior stems from that place. He's delusional and his reality is warped, which is an interesting (and often disturbing) POV to put ourselves in. But a consistently compelling creative challenge."

Meg: "**You wrote our favorite episode in season two of *You*, "Fear and Loathing in Beverly Hills." To us, it felt like it had some influence from *The Shining*. Can you talk about the process for writing that episode in particular?**"

Kara Lee Corthron: "Thank you! And you are absolutely correct. We knew early on that Forty was going to give Joe some kind of hallucinogen in this episode, but all *The Shining* references came during the writing process. Full disclosure: the 1980 movie is one of my all-time faves; long before I was on the *You* staff, I had a T-shirt with the Grady twins on it and I also got another tee as a gift featuring the carpet from the Overlook Hotel (my Gmail profile pic is a still from the film and has been for a few years). When we realized the episode was going to take place in a hotel during the acid trip, we thought about how horrific that experience would be for Joe—a person more likely to murder the one he loves than to give up his sense of control over her. That coupled with the fact that Forty has them imprisoned in order to write a screenplay—not unlike Jack Torrance moving his family to a remote outpost in order to write a novel—sent us in the direction of *The Shining*."

Meg: "We're *Shining* fangirls too and have the Grady girls as tattoos!"

Kelly: **"How do you navigate changing some key plot points when adapting a novel to the screen like the writers did for the series *You*?"**

Kara Lee Corthron: "There are just different considerations. For the show, we have to think about increasing tension and suspense over a set amount of time. For *You*, it's ten episodes a season. We're essentially telling the story in increments as though it were one long novel whereas Caroline has more control when she writes a book, deciding when and where to pick up in her Joe Goldberg story and adding or subtracting characters according to what feels right for the book version of Joe. We have other considerations like casting. For instance, if Caroline decided she wanted to write a book that begins with Joe at age sixty, we wouldn't be able to do that because our lead actor is a crucial part of the show's success and he's a thirty-three-year-old man. Sometimes it also comes down to elements that make good TV that might not be the case for fiction. For example, the Joe/Paco storyline in season one was very popular with some of the higher-ups in our studio and network, which led us to create a similar, but different dynamic in season two with the Joe/Ellie storyline. Neither of those arcs exist in the books, but they worked really well on TV."

Meg: "**Do you have a favorite Stephen King novel or movie/ television adaptation? What is it about the work that you like?**"

Kara Lee Corthron: "As mentioned before, *The Shining*. There is a lot that I love about that movie so I'll just name a few things. I love the messiness. I've seen it more times than I'd like to admit. I've seen *Room 237* (2013) a few times and there are still parts of that movie that make *no* sense to me and I'm totally okay with that. Was Jack Torrance a guest at the Overlook in 1921? Was it Delbert Grady? Does Delbert Grady sometimes look like Jack Nicholson? I have no idea. (You may be an expert with answers to these questions, but I respectfully ask that you don't provide them.) How does Jack get freed from the freezer? Did he and Danny's combined shining powers bring the ghosts to life? Well . . . probably. Regardless, the images, the performances, the score, the humor, and the terror of that film make it a masterpiece in my eyes."

Meg: "It's really interesting to learn about the differences between fiction writing and screenwriting and to think about how calculated adaptations really are!"

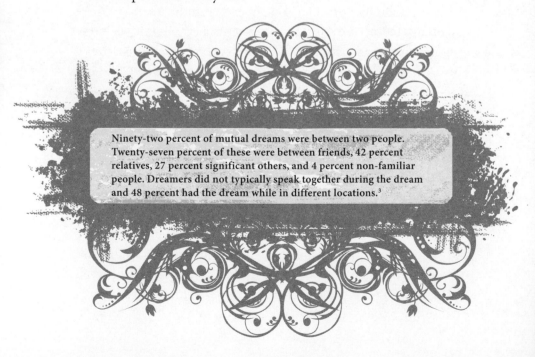

Ninety-two percent of mutual dreams were between two people. Twenty-seven percent of these were between friends, 42 percent relatives, 27 percent significant others, and 4 percent non-familiar people. Dreamers did not typically speak together during the dream and 48 percent had the dream while in different locations.[3]

One theme in *Dreamcatcher* is the idea of shared dreams. Although no scientific studies have been conducted on this phenomenon, there is plenty of anecdotal evidence of people having the same dream. It tends to be more prominent in twins and longtime married couples but is even reported by people who are strangers and recognize each other from their dream. While shared dreams haven't been properly researched, there exists a lot of science around dreams. According to Antti Revonsuo, a psychologist at the University of Turku in Finland, dreams may play a role in helping us deal with possible threats. "His threat simulation theory proposes that night-mares about attacks by saber-toothed cats gave our early ancestors an opportunity to practice how to survive a real-world encounter."[4] Maybe this is why we, as horror fans, like the genre so much! We may be prac-ticing how to handle these seemingly impossible obstacles. Our dreams act as a reservoir of ideas and allow us to process and work through personal feelings, subconscious opinions, and memories.

Dreamcatchers originated in the Ojibwe culture and are traditionally hung over a baby's crib to offer protection.

The character of Duddits in the novel plays a major role throughout the story. He is the reason the four friends come together in their child-hood and again by the end of the book. Duddits has Down syndrome, which is the most common chromosomal condition diagnosed in the United States. There are about six thousand babies born with it per year. Those who have Down syndrome may have some common physical traits of the condition including low muscle tone, small stature, and an upward slant to the eyes. Children with Down syndrome tend to have a higher incidence of infection, respiratory, vision, and hearing problems but can lead healthy lives. The average life expectancy of individuals with Down syndrome is sixty years.

The main culprit causing havoc in *Dreamcatcher* is the strange, reddish fungus left by the alien parasite. When it's inhaled or eaten by people or animals, it causes large, wormlike aliens, called byrum, to infect the host. Byrum could be described as lamprey-like creatures with multiple rows of teeth. Another form of byrus can grow on open wounds and mucus membranes. How do spores of fungus actually affect living things? Fungal spores can be triggers of allergic reactions or can be the cause of infectious disease in humans. An example that could be present in our homes is mold. If the spores are airborne, they present possible respiratory problems for people. Fungi don't only affect humans, though. The majority of plant diseases are caused by fungi with over ten thousand species being recognized as pathogens.[5]

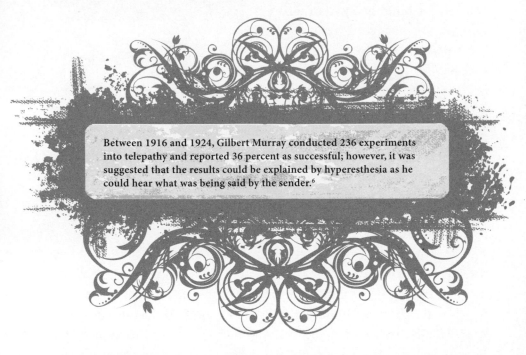

Between 1916 and 1924, Gilbert Murray conducted 236 experiments into telepathy and reported 36 percent as successful; however, it was suggested that the results could be explained by hyperesthesia as he could hear what was being said by the sender.[6]

In the novel, when an infestation is sufficiently established with the byrum, the host develops a form of telepathy with other infected individuals. Do any animals communicate telepathically? Anecdotal evidence suggests that people can communicate with their pets or that their pets have a telepathic sense about certain things. For example, some dog owners claim that their pets are waiting for them when they get home as if they had a sense that they knew they were about to return. Others claim to see animals communicating with each other without using sounds. This could be explained by nonverbal communication between animals. Movement, proximity, and tactile interactions all play a role in conveying messages without words. Proponents of animal telepathy believe humans can communicate with animals if they are open to it.

Several characters and animals have similar symptoms caused by infection with an extraterrestrial macro virus. Army scientists named this "The Ripley," after the protagonist of the *Alien* (1979) movie. It has an extreme resistance to destruction. Do these types of macro viruses exist on Earth? Viruses are entities whose genome replicates inside living cells, and are able to transfer that viral genome to other cells. There are multiple viruses on Earth that are extremely resistant to treatment including Ebola, Hantavirus, and Dengue fever. There could be viruses in space, too, but astrovirology has not been studied much. "Part of the reason for astrovirology's absence from space-science agendas is that virologists have not been reaching out to astrobiologists and pushing the case for virion hunting. Another major reason is technical: virions are tiny and so scientists need transmission electron microscopes to see their unique and varied shapes."[7]

The byrum maintain a symbiotic relationship with their host. Are there examples of this in nature? Some symbiotic relationships can be healthy. For example, bacteria living in our bodies have beneficial effects. In the ocean, sea anemones and hermit crabs help each other by fending off the other's enemies. If the symbiotic relationship between two living things is harmful to one then it is considered parasitic. Examples of this include roundworms, fleas, and barnacles.

Mr. Gray is described as the perfect Typhoid Mary to spread the virus. What is the history behind this infamous figure? Mary Mallon was the first asymptomatic typhoid carrier to be identified by medical science and she worked as a cook for seven families in New York City from 1900 to 1907. During that time, she infected countless people with typhoid fever. Only three deaths are attributed to her but the number could be as high as fifty due to her use of aliases and secrecy. Typhoid is a bacterial infection that causes fever, abdominal pain, headache, and vomiting. Symptoms usually don't present until six to thirty days after exposure so it can be difficult to track the origin. Mallon was eventually quarantined for life and died of pneumonia at age sixty-nine.

In *Dreamcatcher*, Henry, too, is quarantined by the United States Army. How often are quarantines used? In 1799 the first quarantine hospital was built in America after a yellow fever outbreak occurred six years prior. In comparison, the coronavirus outbreak in 2020 prompted China to build a quarantine hospital in only ten days! At one point in the 1970s, the Centers for Disease Prevention and Control reduced the number of quarantine stations from fifty-five to eight because it was believed that infectious diseases were a thing of the past. The number was increased to twenty again within the past two decades due to threats of bioterrorism and other pandemic possibilities.

Although Mr. Gray infects some water supplies in the area, the last of the virus dies in the fire. Written longhand while recovering from his accident, some things remain true; like many Stephen King novels, *Dreamcatcher* ends with a group of friends coming together to defeat the evil. King's own family, friends, and fans certainly showed up for him in his time of need and the world will forever be grateful.

CHAPTER TWENTY-ONE

From a Buick 8

Ubiquitous animals like dogs and birds can seem harmless, even cute. Stephen King harnessed their seemingly placid natures, subverting our expectations of these creatures in novels like *Cujo, Pet Sematary,* and *The Dark Half.* In *Cell*, he reveals the terror of one of our most beloved machines, the cell phone. So, naturally, our four-wheeled companions are not safe from the Master of Horror's gory treatment. This tradition started with the short story *Trucks* (1973). Later adapted into the 1986 Stephen King–directed film *Maximum Overdrive, Trucks* is the story of strangers trapped in a truck stop when the semis outside suddenly get evil minds of their own. Next came the novel *Christine* (1983) in which a haunted 1958 Plymouth Fury causes death and chaos to whomever gets in its supernatural way. Considering the vehicular accident that King endured in 1999, it's no wonder that he revisited murderous cars with the 2002 publication of *From a Buick 8*.

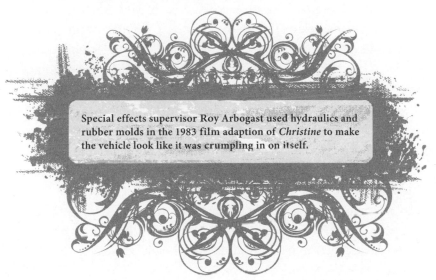

Special effects supervisor Roy Arbogast used hydraulics and rubber molds in the 1983 film adaption of *Christine* to make the vehicle look like it was crumpling in on itself.

Centering on a troop of Pennsylvania cops, the novel unravels the narrative of when the car was abandoned in 1979, and all the chaos it has caused until present day, through multiple perspectives. In fact, it doesn't seem to be a car at all. As described in her review for the *New York Times*, Laura Miller explains the odd nature of the car in question. "Everyone who sees the car instantly knows that it's 'wrong.'

In 1902, Eugene C. Richard applied for a Buick patent on an overhead valve, or valve-in-head, engine. The design was later adapted across the auto industry.

It's got an exhaust system made out of glass and a dashboard full of phony controls, for example—and it just gives people the creeps."[1] The Buick in the novel, described as similar to the make of a Buick Roadmaster, is not just murderous like the Plymouth in *Christine*. It seems to be a portal between worlds; our own, and another that is frightening and vastly different from what the cops are used to. This theme of a schism between two worlds has been explored numerous times in King's work, including in the novella *The Mist* (1980) and the Dark Tower series.

Evil cars are not a new idea created by Stephen King. While he undoubtedly is the horror fiction master, there are urban legends about cursed or possessed cars that stretch further back in modern history. Like the 1964 Dodge 330 that has been blamed for the murder of fourteen people! The story is that the car began as a police vehicle. Three officers who drove the Dodge all committed suicide after murdering their families. It then found its way to Wendy Allen, an owner who claims the doors had a mind of their own. Then came a spate of strange deaths, all happening to people who had vandalized the notorious Dodge. From a strike of lightning to a horrific semi-truck accident, the car sought its bloody revenge. The legend says that it was then chopped into bits by a worried church group. (Okay, there are no documents or proof that this Dodge was truly murderous, but it sure makes for great fiction like *Christine* and *From a Buick 8*.)

A more verifiable story of a possibly "cursed" car is the 550 Spyder owned by movie star James Dean. Nicknamed "L'il Bastard" by Dean, the Spyder caused so much death and destruction, it was purposely hidden from the world. First, it took its famous owner's life. On September 30th, 1955, twenty-four-year-old Dean was killed when he hit a vehicle in rural California. After this tragedy, George Barris, creator of the Batmobile for the Batman TV series (1966–1968) came into possession of the Spyder. He sold a few parts to two doctors who used them in a race. One of the doctors perished. Then, Barris loaned the car out to the California Highway Patrol in order to be used as a deterrent against reckless driving.

> The first place the CHP stored the car was in a garage that promptly burned down, with only the wreckage of the old car left standing. Chalking this up to bad luck, the CHP continued to use the car, taking it to high schools as a visual aid for the dangers of reckless driving. En route to one school, the car broke loose from the truck hauling it and crashed into another vehicle, causing a fatal accident. Undeterred by these bad omens, the CHP took the car to another school, where the car fell on a student, breaking their hip. In total, the Spyder fell off of the trailer that carried it three times, crushing a truck driver once. Not only did the car give law enforcement trouble, but it also made life difficult for criminals. Two thieves tried to steal the bloodstained seats and steering wheel from the wreck. Instead of getting some memorabilia out of it, all they got were injuries.[2]

The mystery of James Dean's Spyder, which went missing shortly after its supposed reign of terror, came back to the headlines in 2015 when forty-seven-year-old Shawn Reilly revealed that as a child he had helped his father hide a sports car behind a wall. During the task, Shawn sustained a scar on his thumb that still lingers today. The Volo Auto Museum in Illinois conducted a lie-detector test on Reilly to substantiate claims that he and his father, along with George Barris, hid the Spyder.

He passed the test.

Whether cars can be possessed or cursed, they are undoubtedly an integral part of our lives. As technology rockets ahead, the reality of self-driving cars has come to be more than just fiction. When Stephen King wrote *Trucks* in the seventies, he couldn't have guessed that in a few short decades cars would, indeed, be driving themselves. In the last decade, the development of self-driving cars, also known as autonomous vehicles, has exploded. Each state has developed safety laws, yet they have changed rapidly, as they are attempting to keep up with technology.

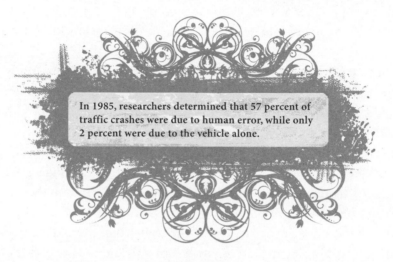

In 1985, researchers determined that 57 percent of traffic crashes were due to human error, while only 2 percent were due to the vehicle alone.

The US Department of Transportation first released guidelines in 2016 to give direction to developers of self-driving technology, and has updated them three times since. The guidelines emphasize safety, tech neutrality, and regulatory consistency, and they encourage companies to submit information about their approaches to safety. But they didn't create any rules around testing self-driving cars. Today, states oversee testing and have taken wildly different approaches. California, for example, requires developers to obtain a license to test, and to annually submit detailed information on its testing activities, which is released to the public. (Many in the industry argue that the data collected by the state doesn't provide a clear or accurate picture of its activities.) Arizona, by contrast, only requires companies to submit information about how their vehicles will interact with law enforcement and emergency services, and to guarantee in writing that their vehicles are autonomous. After that, Arizona does not monitor the vehicles, though its governor reserves the right to revoke any company's ability to operate after something goes wrong. Governor Doug Ducey did just that in 2018, kicking Uber's self-driving operations out of the state after a testing vehicle struck and killed a woman on foot.[3]

Forty-nine-year-old Elaine Herzberg died when she was struck by a self-driving Uber with a human operator in the driver's seat. According to the National Safety Traffic Board, her tragic death was blamed on the vehicle's inability "to classify an object as a pedestrian unless that object was near a crosswalk."[4] Therefore, because it couldn't recognize Herzberg as a person since she was jaywalking, the Uber concluded she was a bike or an "other" and braked only one second before hitting her. Unfortunately, the operator inside the vehicle was not alerted by the vehicle, and was streaming TV on her phone rather than watching the road. This mix of human and technological error cost Elaine Herzberg her life. Because of this accident, Uber has programmed their recent cars to be able to recognize jaywalking humans.

This fatality is listed as the only "level three" autonomous vehicle deadly accident. This refers to the type of car involved. Because the Uber car was autonomous and required only little human intervention it is considered a level three autonomous vehicle (AV). As of September 2019, there have been five worldwide fatalities due to "level two" self-driving cars. All were Tesla models. These require more human involvement. It's interesting to make note that the more people were expected to intervene, the more deaths occurred. Ranging from 2016–2019, one of these deaths happened in China, while the other four were in the United States. The first death was twenty-three-year-old Gao Yuning who after turning on the AV feature in his Tesla, was slammed into the back of a truck. In Florida, Joshua Brown was killed when his Tesla's AV function mistook a left-turning white truck for clear sky. While Tesla has come under fire for such tech mishaps, they point to statistics to prove that self-driving cars are safe, if not safer than the average car. In a letter to the public after Yuning's death they emphasized the point: "This is the first known fatality in just over one hundred and thirty million miles where Autopilot was activated. Among all vehicles in the US, there is fatality every ninety-four million miles. Worldwide, there is a fatality approximately every sixty million miles.[5]

With the advent of AVs, there comes ethical questions. First are the legal implications. Is the human operator to blame if the vehicle did not alert them to a problem? Is the manufacturer solely responsible for fatalities? Beyond these considerations are questions revolving around artificial intelligence.

The Trolley Problem is an ethical dilemma often posed to philosophy students. The scenario is this: If you were to stand beside a lever that controls the direction of a trolley that cannot stop, and there are five people tied on the tracks, would you divert the trolley seeing that there is one person on the other track? Would you essentially jump into action to kill one person over five? What if the five were children? Or the single one? There are many deviations of the scenario, yet they ask the same fundamental ethical questions. The Utilitarian approach would be to divert the trolley, because one death is preferable to five, yet there is a differing opinion in which participating in the carnage is morally wrong.

For example, examining how the technology might respond to a "no win" situation or a true ethical dilemma, such as the Trolley problem, is a vital task. Some scholars are seeking to evaluate

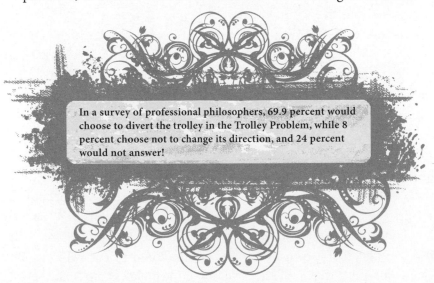

In a survey of professional philosophers, 69.9 percent would choose to divert the trolley in the Trolley Problem, while 8 percent choose not to change its direction, and 24 percent would not answer!

empirically how drivers might handle a Trolley-type of situation; others are considering whether ethical theory, such as the application of Utilitarian reasoning may help resolve what is an appropriate response by the car in a crash situation."[6]

Whatever you believe the right answer to be, there is worry that self-driving cars will never have the human ability to make these ethical choices, particularly in a matter of seconds. Many inanimate objects scare us, from dolls to cars. In *From a Buick 8*, the murderous car seems to protect itself by devouring those who get too close. If cars were sentient, would it be so hard to believe they would work to preserve themselves, like any human? While Stephen King writes about horror in many different forms, it's important to see that it is not the fictional creatures, but rather old-fashioned people who cause the real chaos, as King himself said. "Evil is inside us. The older I get, the less I think there's some sort of outside devilish influence; it comes from people. And unless we're able to address that issue, sooner or later, we'll fucking kill ourselves."[7]

CHAPTER TWENTY-TWO

Cell

It's not so odd these days to see groups of people all on their cell phones at the same time. Look at passengers on a subway or a train and most of them will be interacting with their device in some way. This rampant surge in the availability and popularity of technology got Stephen King thinking. *What would happen if our cell phones were used as a weapon against us?*

> While walking on the sidewalk in New York, Steve saw a man dressed in a business suit approaching him and seemingly engaged in conversation with himself. Since people talking to themselves on a public street would usually be among those whose "reality" was a bit different than the norm, Steve became a bit apprehensive. He realized as the man got closer, however, that he was talking on a cell phone using a headset. It was the incongruity of a person who may not be in touch with reality, but dressed in a business suit that sparked the idea for *Cell*."[1]

Cell phones in the novel *Cell* are the devices that incite violence among people. How long have phones been around? The telephone was the result of the work of many people and is most often attributed to Alexander Graham Bell, who patented the device in 1875. The road to the modern telephone, though, has a long and complicated history. It began in 1667 when Robert Hooke invented a string telephone that conveyed sounds over an

The telephone was invented in 1875 and patented by Alexander Graham Bell.

extended wire by mechanical vibrations.[2] This is similar to a tin can phone that many children played with in generations past. Ironically, Robert Hooke also coined the term "cell" in relation to microscopy! (Not in relation to the telephone.) In 1861, Philipp Reis constructed the first telephone that transmitted speech. Using his knowledge of physics and electricity, he constructed a device that was able to send messages up to a distance of a little over three hundred feet. The invention didn't gain much interest in his home country of Germany, though, and he didn't gain recognition for his contributions to the technology until after his death. It wasn't until 1927 that the first transatlantic phone call took place. The call was placed from New York in the United States to London in the United Kingdom. Listening to the audio, you can hear clicks and static with more than a few "hello's" spoken to gauge if the listener is present on the other end. A short speech is made proclaiming business can now happen between the two places almost instantly. How far we've come in less than one hundred years!

In 1973, ten years before a cell phone was first released onto the market, the first cell phone call was made by Motorola researcher and executive Martin Cooper.[3]

"The Pulse" is sent out across cell phones in the novel and turns people into mindless killers. Could sound be used as a weapon? Sound waves, when used with high power, are able to unsettle or destroy the eardrums of the intended target. Even less powerful waves have the effect of causing nausea in humans and can cause discomfort. A long-range acoustic device can send messages over longer distances that a normal loudspeaker wouldn't be capable of. This device has also been used to deter crowds, rioters, and criminals over the past decade. Noise is even more powerful than that, though. Studies have shown that exposure to high intensity ultrasound at frequencies from 700 kHz to 3.6 MHz can cause lung and intestinal damage in mice.[4] Sonic weapons aren't a new invention, though. It is rumored that in World War II, Adolf Hitler's chief architect, Albert Speer, set up a lab in order to explore his theories of sonic warfare. He created an "acoustic cannon" that worked by igniting a mixture of methane and oxygen in a resonant chamber. In theory, it could create a series of over one thousand explosions per second. This, in turn, could kill someone standing within a one-hundred-yard radius in as little as thirty seconds. This weapon was never actually used, but the consequences of its possible existence are terrifying.

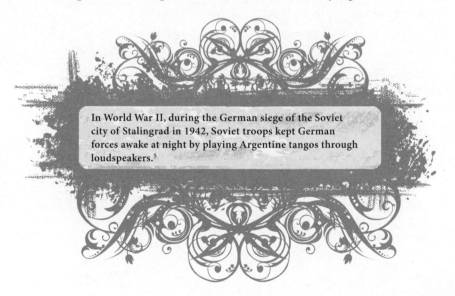

In World War II, during the German siege of the Soviet city of Stalingrad in 1942, Soviet troops kept German forces awake at night by playing Argentine tangos through loudspeakers.[5]

One thing unique to "The Pulse" victims is that they foam at the mouth. We most often associate this phenomenon with a dog suffering the effects of rabies but is it possible in humans? It is, although it is rare. The most common causes of foaming at the mouth are a drug overdose, seizure, or rabies. Foam is produced when saliva mixes with air or other gases. Because it is so rare, it's recommended that you seek medical attention immediately if you experience it.

There are a lot of ways we try to conserve our cell phone batteries in daily life and one is used in the movie version of *Cell*. A character recommends putting their cell phones in the freezer in order to make the batteries last longer. Is this a good idea? According to experts, it's not recommended. Cold temperatures can harm batteries especially if condensation occurs when bringing the cell phone or battery back to room temperature.[6] To save cell phone battery, it's best to adjust settings on your cell phone. Delete or shut down apps that use the most battery, turn off location services, and turn on the low power mode setting if available. These are good everyday tips to save some battery life but can also come in handy if you're on the run from some creatures like the characters in this novel!

Technology's effects on living things is a theme in *Cell* that hasn't been fully explored by experts. Since cell phones and other devices are relatively new inventions, what do we know about how these things affect us? Some things may be obvious. The more time we spend on devices, the less we may be physically active. This could ultimately affect our health, and even eyesight, but isn't necessarily unique to technology itself. A bit more nefarious are the effects that cell phone radiation may have on living things. According to a study conducted in 2010, "the indiscriminate use of wireless technologies, particularly of cell phones, has increased the health risks among living organisms including plants. The study concludes that cell phone EMFs impair early growth of V

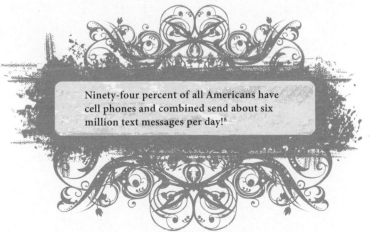

Ninety-four percent of all Americans have cell phones and combined send about six million text messages per day![8]

radiata seedlings by inducing biochemical changes."[7] The effects of having a cell phone in close proximity to your brain are not yet conclusive. It is known that radiation can be harmful but the long-term effects are not likely to be understood for a long time to come.

The people in the book move a bit like zombies but even more like birds. How do birds move in the way that they do? It's quite a sight to see when you gaze upon a large group of birds flying in the sky together and they instinctively and automatically make a quick, calculated turn. This seemingly miraculous movement had even ancient Romans believing that the gods were speaking through flocks of birds. More recent scientists speculated that birds could communicate through telepathy or a type of biological radio. Current observations with photography and computer models have learned that the movement of birds comes down to natural, quick instincts. Though not visible to the naked eye, birds follow the movement of the bird in front of them when flying in a flock. Think of it as a giant wave in a football stadium except quicker. Birds are able to pay attention to their closest six or seven neighbors in a flock and follow their movements precisely.[9] This helps them to evade predators and to move in sync. Researchers agree that they don't yet know everything about how birds move in groups but speculate that they could be using auditory or even tactile communication in order to analyze space and movement. Unlike the affected humans in *Cell*, birds seem a lot more intelligent!

Although Stephen King's novel *Cell* may seem dated in some ways, humanity's link to technology is certainly still relevant. If the threat of a pulse that could control much of Earth's population became a threat, could we put our devices down? Hopefully. But if not, this book gives us an idea of how we could fight to survive if we are one of the only ones left.

CHAPTER TWENTY-THREE

Lisey's Story

In the summer of 1999, Stephen King was walking in the rural backroads of Lovell, Maine, for a bit of fresh air and exercise. Local resident Bryan Smith drove a van on the county road, and was purportedly distracted by a dog in his backseat, when he struck King from behind, sending him fourteen feet from the pavement. King was immediately brought to a nearby hospital, and then, because of his serious injuries, airlifted to the Central Maine Hospital. This watershed moment in Stephen King's life has forever altered his fiction. A character named "Brian Smith" appears in *The Dark Tower VII: The Dark Tower* (2004), in which he is high on drugs and distracted by a dog in his vehicle. This theme also pops up in the TV series developed by King, *Kingdom Hospital* (2004).

In a more meaningful way, the incident inspired him to write *Lisey's Story*, a novel unlike most of his body of work. While *Lisey's Story* has a King-esque paranormal element, it is at its heart, a story about marriage and grief. While Stephen King was in the hospital, his wife, Tabitha, organized his office. When he returned, after a near-death experience, he was struck by what his office would look like if he had lost his life on that summer day. When asked in an interview if *Lisey's Story* is about his own mortality, King was quick to admit it. "Sure, there's no doubt about that. I had the accident, and then as kind of an outfall of the accident, two years later I had pneumonia because the bottom of my right lung was crumpled and nobody realized that. It got infected and that was very serious, that was actually closer [to death] than the accident. So, I had some of those mortality issues."[1]

In *Lisey's Story*, Lisa Landon's famous author husband is not so lucky. He dies, leaving her with just memories. As she navigates through her grief, Lisa must face the reality that Scott had been escaping to an alternate reality their entire marriage, a place called Boo'ya Moon. Desperate

From the car accident, Stephen King suffered broken ribs, a broken hip and leg, and a punctured lung. He spent three weeks in the hospital and endured five surgeries.

to reconnect to her love, Lisa wants to traverse this beautiful yet frightening place to find him. This led us to question the vast depths of grief. And what better way than speaking with someone who is well-versed in this difficult topic? In order to learn more about the reality of grief we spoke to licensed professional counselor, and Stephen King fan, Samantha Dansby.

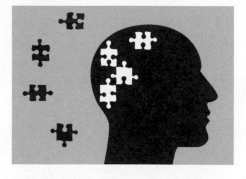

Repressed memories refer to the rare psychological phenomenon in which memories of traumatic events may be stored in the unconscious mind and blocked from normal conscious recall.

Meg: **"First, can you tell us about your professional background?"**
Samantha Dansby: "I have been working as a counselor since 2015. I've worked in a variety of settings, including an inpatient psychiatric hospital for teenagers, a private practice, a clinic that provided pro-bono counseling services, and for a community mental health organization. I've worked with a variety of mental health disorders from mild to severe and with clients ranging from the age of one to eighty-five."

Kelly: "**Wow! That's quite a resume! What led you to pursue becoming a counselor?**"

Samantha Dansby: "I have struggled with my own mental health throughout most of my life, as well as watching many others I love struggle. From the time I can remember, all I really wanted to do was help others, and counseling gave me that outlet. I lost my mother at the age of twenty-two and my father at the age of twenty-six, so that pushed me more to want to work with grief specifically."

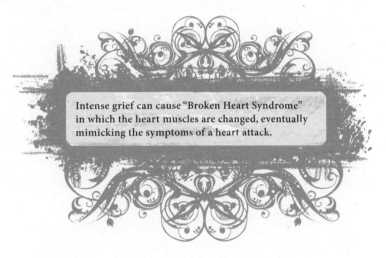

Intense grief can cause "Broken Heart Syndrome" in which the heart muscles are changed, eventually mimicking the symptoms of a heart attack.

Meg: "**I'm sorry to hear that. How would you describe the grieving process? Is it unique to every human? Do there seem to be shared experiences?**"

Samantha Dansby: "Oh, wow. How to describe the grieving process? I honestly think that the grieving process is a journey. I don't know that it is a journey that ever ends. My mom died eight years ago, and I'm still grieving. One of the best comparisons I have heard is comparing grief to an ocean. At times, the waves are calm, and at times, they crash. All we can do is find some way to keep swimming. There are many models of grief in the counseling world, but I truly do not believe that it is a step-by-step process.

I also don't think anyone can look at anyone else who is grieving and honestly say, 'I understand.'"

Kelly: **"You have counseled many through their grief. In the King novel *Lisey's Story*, Lisey Landon is navigating through the loss of her husband as she handles getting rid of his books and keepsakes. From your experience, is this a vulnerable time for those grieving?"**

Samantha Dansby: "Absolutely. Both through my personal experience and through the accounts of my clients, I've seen the struggles of trying to sort through the possessions of a person, of realizing that those things are all that physically remain of someone you love. Often, people are scared to part with anything for fear that they might lose even more of the person that is gone."

Meg: **"Lisey begins to see, hear, and even feel the presence of her deceased husband, which King explains with the supernatural parallel universe called Boo'ya Moon. But in reality, is it common for people to experience auditory or visual hallucinations of their lost loved one while in extreme grief?"**

Samantha Dansby: "Honestly, this varies wildly from person to person. I've had clients who believed that their loved ones still visit them, sometimes in dreams and sometimes when they are awake. I've had other clients who lamented that they could never feel the presence of their loved one once they had passed. I believe most of it goes back to a person's deeply held beliefs about life, death, and the afterlife. I think people who are raised to believe in an afterlife and in a spiritual realm are more likely to see their loved ones after death."

Kelly: **"As a child, precocious and creative Scott Landon finds solace in the alternate world, Boo'ya Moon, due to his troubled childhood. Have you dealt with patients who have dissociated, or created their own reality? Why does the human brain employ this tactic?"**

Samantha Dansby: "Dissociation is such a fascinating subject. I took an intense interest in it when I was a student and wrote numerous papers about it, and I continue to study it whenever I get a chance. According to the National Alliance on Mental Illness (NAMI), approximately 75 percent of people experience at least one episode of dissociation, although only about 2 percent of those people are diagnosed with a dissociative disorder. The trauma we experience is sometimes too much for us to bear, so our brains compensate to protect us by dissociating. We may begin to feel as if the real world is not real, almost as if it is a movie or TV show we are watching from the outside. I believe Scott created Boo'ya Moon to cope with the severe abuse by his father and retreated further there to deal with the traumatic death of his brother, Paul."

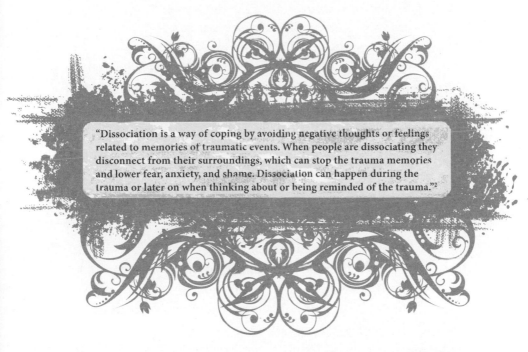

"Dissociation is a way of coping by avoiding negative thoughts or feelings related to memories of traumatic events. When people are dissociating they disconnect from their surroundings, which can stop the trauma memories and lower fear, anxiety, and shame. Dissociation can happen during the trauma or later on when thinking about or being reminded of the trauma."[2]

Meg: "**When you watch movies or read books with fictional depictions of counselors like yourself, what do they get right? Are there false ideas about your profession that you would like to correct?**"

Samantha Dansby: "There are so many frustrations on my part when I see fictional depictions of counselors. It is bad enough that most people don't want to even watch anything with a counselor in it with me. First off, no one has ever lain down to talk to me. I don't even own a chaise! When I worked with kids, most of our sessions took place on the floor or on my beanbags. With adults, most of them sit on a couch or in a chair. I'm also not sleeping with any of my clients. Also, I don't give advice. I listen and help others come to their own decisions about what is best for them. Counselors are more of a sounding board than anything else. Sometimes, I share techniques that could be helpful, but I never tell someone what they should do. That's not my job. As far as what the media gets right, counselors often do struggle with our own mental health. I know many counselors who are either currently in counseling or have been in the past. Another correct assumption, at least on my part, is that it is extremely hard to leave work at work. Empathy is a skill essential for a good counselor, and it is not something I can just turn on and off on a whim. I have to be intentional about my self-care and making sure I take the time to decompress after work."

Kelly: **"We heard you met your best friend in an interesting way? Also, what is your favorite Stephen King work and why?"**

Samantha Dansby: "I love sharing this story! I've got Stephen King to thank for my best friend. We are from a very small town in rural Alabama. One of my favorite places growing up was the library. Once I discovered Stephen King, who took up two whole shelves, I was hooked! It became my favorite area in the library, and I would browse those two shelves, reading book jackets, and savoring choosing my next book. One day in the summer before ninth grade, I rounded the shelf and nearly ran into a girl close to my age who was browsing the SK books. We started chatting and soon realized our shared horror and true crime obsessions. That was over seventeen years ago, and she is still absolutely my best friend. And, as for my favorite SK book . . . how dare you ask me such a thing! But truly, that is an extremely hard question for someone like me to answer. My first book by him was *The Bachman Books*, and *Rage* was a story that stuck with me deeply, but I would not call it my favorite. Over the years, my favorites by him have been *Salem's Lot, Under the Dome, and Needful Things*. I loved *Salem's Lot*, as vampires have always been one of my favorite creatures. *Under the Dome* just captivated me in some way, although I can't really say why. As far as *Needful Things* goes, I have always loved thrift stores, and I often think of this book when I am visiting a new one. Plus, who isn't always on the lookout for the perfect treasure? However, *The Stand* is next on my TBR list, so who knows what my favorite might be next?"

Meg: "There are so many Stephen King books to discover!"

Thank you to Samantha for her insights on grief. It is a difficult conversation, but one that touches every human.

While Boo'ya Moon is a real place in *Lisey's Story*, it is not difficult to understand that we are often driven to live in a fantasy world when faced with grief, abuse, or loneliness. Perhaps that is why Stephen King's work resonates. He creates worlds we can get lost in.

CHAPTER TWENTY-FOUR
Duma Key

Stephen King's accident in the summer of 1999 was partially responsible for another novel. In *Duma Key*, the protagonist, Edgar Freemantle, a contractor in Minneapolis, has an event that forever changes his life: A construction crane falls on him leaving him with a fractured skull, a shattered hip, and an arm that needs to be amputated. "Edgar's injuries were worse than mine," King recalled in an interview with *USA Today*. "I didn't lose an arm, I didn't lose my wife, but like him, my memory was affected. I know a little about pain and suffering and what happens when the painkillers lose their efficacy, when your body gets used to them."[1] The novel is the first that's set in Florida and Minnesota.

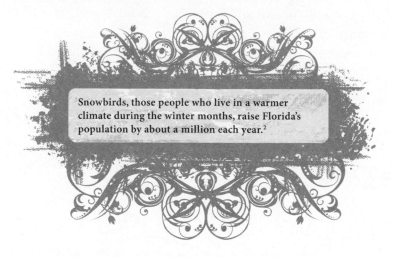

Snowbirds, those people who live in a warmer climate during the winter months, raise Florida's population by about a million each year.[2]

How common is amputation currently in medicine? There are nearly two million people living with limb loss in the United States and among those, the main causes are vascular disease (54 percent), trauma (45 percent), and cancer (less than 2 percent).[3] Approximately 185,000 amputations occur in the United States each year. Amputations date back to the time of Hippocrates and were performed most often due to injury or war. With loss of blood, though, amputations were often dangerous and not necessarily lifesaving. Not until the 1500s did physicians begin using ligatures to restrict a patient's blood vessels. Tourniquets were introduced in 1674 and continued to be used during the period of World War I, which saw an unprecedented number of amputations.

Edgar experiences phantom limb sensation due to his amputation. Is this common? It was first reported in 1551 by surgeon Ambrose Pare. He noticed that patients would complain about pain or a sensation in a limb after it had been removed. Studies show nearly all people who undergo an amputation experience this phenomenon. Why is that? Understanding the way our brains work, phantom limb sensations could be due to reorganization in the somatosensory cortex. This part of the brain receives and processes sensory information for the entire body including the sensations of touch, pain, and vibration. Doctors don't agree on exact causes for this phenomenon but treat it using medications, hypnosis, and biofeedback. Mirror therapy, in which a person views their working limb in a mirror and mimics the movement with their phantom limb, is sometimes used but has not proven to be entirely effective.

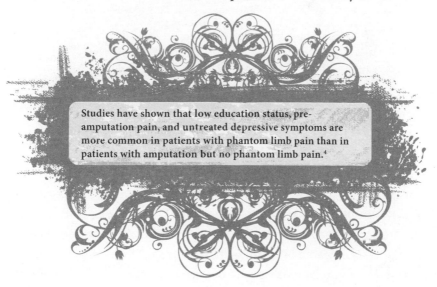

Studies have shown that low education status, pre-amputation pain, and untreated depressive symptoms are more common in patients with phantom limb pain than in patients with amputation but no phantom limb pain.[4]

Edgar suffers a traumatic head injury after his accident. His moods become volatile and he and his wife get divorced. How can a brain injury change a person's personality? Brain injuries can damage connections that go from the cerebral cortex to the limbic system. The cerebral cortex is the part of the brain that affects memory, attention, perception, cognition, and awareness. The limbic system is the part of the brain that has to do with emotions, behavior, and motivation. When everything is in working order, we are able to evaluate our emotional reactions and respond in a reasonable way. When these connections are damaged in some way, we may react differently. The most famous case of someone having a changed personality after a brain injury is Phineas Gage. He was a railroad worker who survived having an iron rod go through his head. His personality and behavior changed so significantly afterward that his friends said he was no longer himself.[5]

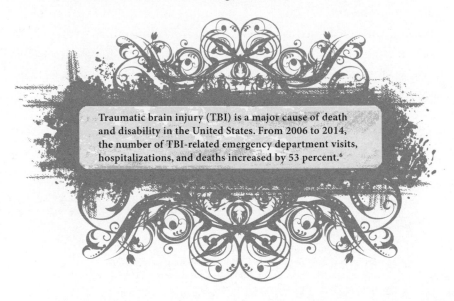

Traumatic brain injury (TBI) is a major cause of death and disability in the United States. From 2006 to 2014, the number of TBI-related emergency department visits, hospitalizations, and deaths increased by 53 percent.[6]

In *Duma Key,* Edgar's doctor recommends he move to Florida for some "geographical healing." Does weather affect mood and healing? According to experts, absolutely! Edgar is encouraged to move out of cold, snowy Minnesota to warm, sunny Florida. Those of us who live in Minnesota absolutely understand how the lack of warm weather and the opportunity to be outside can affect mood. Seasonal Affective Disorder

(SAD) is a real condition that affects people's mood. Being exposed to sunlight, especially during morning hours, can greatly change someone's outlook for the day. Warmer weather has also been linked to people self-reporting that they are happier and have less stress.[7]

Even physical wounds can be impacted by weather. According to The Wound Care Education Institute, cold weather can negatively influence the healing process and may affect how you care for and dress your wounds.[8] Research has found that those in warmer climates are more likely to stay physically active after an injury and will have increased blood flow and circulation. Cold weather could also cause skin or wounds to dry out or become chapped whereas wounds that are kept moist heal 50 percent faster.

Edgar begins to paint when he's in Florida and it seems to be helping with his mood. To understand more about art therapy, we spoke to Theresa Hoglund Mueller, an art therapist who works for Creative Pilgrimages.

Art therapy can help people reduce levels of stress, anxiety, and burnout connected to their work or everyday life.

Kelly: **"How did you become interested in art therapy?"**

Theresa Hoglund Mueller: "I have a strong background in both art and in human services. I have also always been fascinated with the 'flow' state that art-making involves which brings people to 'center,' if you will. Also, the symbolism of dreams fascinated me. I never met one of my grandfathers but I knew him very well through his art, including a series of humorous yet gory cartoons he drew to cope with his stomach cancer that killed him."

Kelly: "Wow! I'd love to see those."

Meg: **"What training do you need to become an art therapist?"**

Theresa Hoglund Mueller: "It requires a clinical master's degree with an intensive internship of about eight hundred hours and post-supervision work of a thousand hours. Prerequisites for the masters include specific undergrad work in psychology and art classes and proving competence in a variety of mediums with a submitted portfolio. Credentialing only occurs after the post-supervision work starting with registration and after passing an intensive exam to prove competence and skills for those who are board certified."

Kelly: **"In Stephen King's book *Duma Key,* the main character begins painting and starts to feel better mentally and physically. How can art therapy improve a person's mood and/or physical health?"**

Theresa Hoglund Mueller: "Doing art can be both very cathartic and meditative given the right application. It can help someone in getting in a flow or altered state of mind. In such states, the mind can stop racing thoughts and the body can begin to relax. A person becomes ready to problem solve in this more relaxed, less judgmental state."

Meg: **"How have you seen art therapy impact people?"**

Theresa Hoglund Mueller: "I have seen profound catharsis, needed calming, and much insight from my clients allowing them to see themselves in new ways that allow them to be open to needed change and/or acceptance."

After speaking with Theresa Hoglund Mueller, it becomes clear that art therapy can have an immense impact on people as it did with Edgar in the book. Although, not in the supernatural way!

Elizabeth Eastlake, a character that lives on Duma Key, has experienced much of what Edgar is going through. She has Alzheimer's disease, though, and this causes Edgar to not trust the things she is saying initially. How does Alzheimer's affect the brain? In a healthy brain, neurons communicate with each other and transmit information via electrical and chemical signals. The brain then sends messages to the rest of the body including muscles and organs. Alzheimer's gets in the way of this communication and, in turn, causes cell death and loss of function. People with this disease lose their ability to function and live independently and will eventually die from it.

Elizabeth, like Edgar, believes that the art created on Duma Key can change the fate of the subjects painted. Could they be suffering from a shared delusion? According to research, most people who suffer from folie-a-deux, or a madness shared by two, tend to have a strong emotional connection or family ties. They also tend to be isolated in some way either geographically or culturally. One study revealed that even a family dog shared in the delusions! This occurrence was documented in *The American Journal of Psychiatry*:

> Ms. A, an eighty-three-year-old widow who had lived alone for fifteen years, complained that the occupant of an upstairs flat was excessively noisy and that he moved furniture around late at night to disturb her. Over a period of six months, she developed delusionary persecutory ideas about this man. He wanted to frighten her from her home and had started to transmit "violet rays" through the ceiling to harm her and her ten-year-old female mongrel dog. Ms. A attributed a sprained back and chest pains to the effect of the rays and had become concerned that her dog had started scratching at night when the ray activity was at its greatest. For protection, she had placed her mattress under the kitchen table and slept there at night. She constructed what she called an "air raid shelter" for her dog from a small table and a pile of suitcases and insisted that the dog sleep in it. When I visited Ms. A at her home, it was apparent that the dog's behavior had become so conditioned by that of its owner that upon hearing any sound from the flat upstairs, such as a door closing, it would immediately go to the kitchen and enter the shelter.[9]

Although Edgar and Elizabeth could be suffering from a shared delusion, they realize that there truly is a villain that is trying to control them.

Jerome Wireman, the live-in attendant for Elizabeth, is grief stricken over the deaths of his wife and daughter. He is so depressed that at one point he attempts to die by suicide. What is the psychology behind grief? There are five stages of the grieving process according to Elisabeth Kubler-Ross in her book *On Death and Dying*:

1. Denial.
2. Anger.
3. Bargaining.
4. Depression.
5. Acceptance.[10]

People tend to spend various lengths of time in these stages and each one may affect others differently. For example, some people may grieve a loss more before an actual death. This is referred to as anticipatory grief. The subsequent passing, then, can sometimes be seen as a relief and the acceptance stage is reached more quickly. One psychologist found that 50–60 percent of people recover from grief fairly quickly while only 10 require therapy or other interventions to help them cope.[11] However you grieve, it's seen as a personal and necessary process that is different for everyone.

Edgar uses his power one final time at the end of the novel *Duma Key* to destroy the island. He returns home and traps the villain in a statuette, dropping it into a freshwater lake in Minnesota. His geographical healing has come to an end and the true healing can begin.

CHAPTER TWENTY-FIVE
Under the Dome

As the Coronavirus pandemic rocks the world in 2020, it would be difficult not to compare the isolation in Stephen King's forty-eighth novel, *Under the Dome*, to this modern crisis. Published in 2009, *Under the Dome* centers on the town of Chester's Mills, Maine; a town that is inexplicably covered with a glass dome. This barrier causes damage and death when it falls from the sky, and more, it causes the townspeople of Chester's Mills to be cut off from the outside world. As the novel progresses, people become more desperate to escape their cage, especially as resources dwindle and the air grows toxic. Eventually, we learn that aliens, known as "leatherheads," are to blame for the dome. In the tradition of many extraterrestrials in the media, they find entertainment in torturing humans, watching as they panic, pillage, and kill. King explained his thought process behind *Under the Dome*:

> From the very beginning, I saw it as a chance to write about the serious ecological problems that we face in the world today. The fact is, we all live under the dome. We have this little blue world that we've all seen from outer space, and it appears like that's about all there is. It's a natural allegorical situation, without whamming the reader over the head with it. I don't like books where everything stands for everything else. It works with *Animal Farm*: You can be a child and read it as a story about animals, but when you're older, you realize it's about communism, capitalism, fascism. That's the genius of Orwell. But I love the idea about isolating these people, addressing the questions that we face. We're a blue planet in a corner of the galaxy, and for all the satellites and probes and Hubble pictures, we haven't seen evidence of anyone else. There's

nothing like ours. We have to conclude we're on our own, and we have to deal with it. We're under the dome. All of us.[1]

This humbling realization, that we are likely alone in the universe, is what makes the parables in *Under the Dome* so powerful. No matter how big or small our personal "bubble," we eventually have to come to terms with our own human vulnerability.

But what about the vulnerability of the planet? When an explosion kills hundreds of residents in *Under the Dome*, the toxic air lingers, not able to escape the barrier. If Earth is, indeed, our shared dome, how have humans destroyed it, both unknowingly and brazenly?

Global warming is a scientific term that has been gaining popularity in the last several decades. It refers to the proven increase of Earth's temperature caused by humans. While global warming has occurred in the prehistoric past, the rise in the twentieth century is the most markedly dramatic. This is due to carbon emissions. Since the

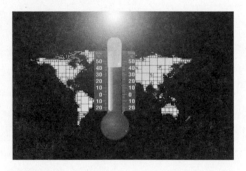

Atmospheric CO_2 has increased since the Industrial Revolution.

Industrial Revolution in the eighteenth century, humans have increased our carbon emissions into the atmosphere by 45 percent. This is because of our use of fossil fuels, as well as extreme man-made changes in the planet, like deforestation. In a study focusing on the US mid-Atlantic region, researchers pinpoint the problems with increased carbon (CO2) emissions:

The Intergovernmental Panel for Climate Change (IPCC) indicates that an average global warming of about 2°C is likely by the end of the century, based on medium- to high-emission scenarios. Along the US East Coast, the warming is projected to be even stronger with an increase of 2.5°–5.5°C. Heat waves are expected to be more frequent and more intense and to last longer. For example, the number of days above 90°F (32°C) may increase by fifty days in the mid-Atlantic region by the middle of the century if CO2 emissions continue on their current trajectory. Such increases in extreme temperatures are likely to have a negative impact on public health all along the mid-Atlantic, as high temperatures are associated with increased mortality. This temperature–mortality link appears to be strongest in the eastern United States compared to other parts of the country.[2]

While there are some who still choose to deny the threat of global warming, scientists can attest that not only is the data relevant, but signs of climate change are all around us.

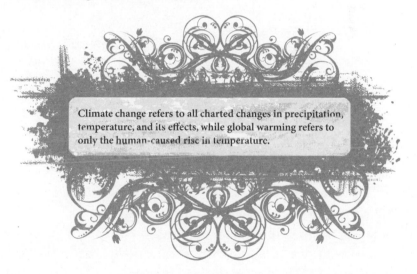

Climate change refers to all charted changes in precipitation, temperature, and its effects, while global warming refers to only the human-caused rise in temperature.

This is most evident in the Arctic, where the temperature has risen the highest. This has caused glaciers and sea ice to shrink. While disappearing ice floes may seem like a faraway issue that doesn't affect the average human, global warming is also the cause of drought, wildfires, and the lessening of crops.

While the future of Earth may sound bleak, many scientists, engineers, and others have devoted their careers to slowing the heat of our very own "dome." This is known as climate change mitigation, and is a crucial development in the fight against global warming. So how can we reverse, pause, or at least slow climate change? First, we must turn to fossil fuel. It is the cause behind 70 percent of greenhouse gases. Thankfully, humans have invented alternate ways to provide energy to their homes and cars with the recent rise in air, solar, and electric technology.

The Australian wildfires of 2019 burned eighteen million acres, killed twenty-nine people, and severely affected the local wildlife.

There are also carbon sinks. These are defined as natural areas that absorb more carbon gases than they release, which, in turn, reduce emissions. The most notable carbon sinks are the ocean and particularly abundant areas of vegetation. While untouched nature is helpful to sustaining our temperatures, researchers have published studies proving that human interference in climate change mitigation is vital. When comparing forests in Europe that are left to grow on their own and their wood taken with ones that are managed for reforestation, there is a notable difference. "The regional climate change mitigation potential of sustainably managed forests is about ten times as high as that of forests taken out of management, based on the lifetime of trees under unmanaged conditions. The difference is mainly due to the substitution effect from the use of discarded wood products as feedstock for bioenergy."[3]

Like the people of Chester's Mills, we did not ask for our "dome," yet it is our responsibility to take care of it. As teenage activist Greta Thunberg has said, "the climate crisis has already been solved. We already have the facts and solutions. All we have to do is wake up and change."[4] So, what can the average person do to reduce their carbon footprint? While giving up driving a car altogether would be ideal, the reality is we all need transportation. There are ways to drive but still be conscientious; consider carpooling, reducing your use of air-conditioning, and using cruise control on long drives. Flying less is also a way to significantly reduce your personal emissions.

Eating less meat is another proven way to create less demand for livestock, thus lessening carbon emissions. According to a study published in 2017 in the journal *Environmental Research Letters*, red meat can have up to a hundred times the environmental impact of plant-based food. According to some estimates, beef gives off more than six pounds of carbon dioxide per serving; the amount created per serving by rice, legumes, carrots, apples, or potatoes is less than half a pound.[5] While you are tweaking your diet, take a look at the appliances in your kitchen. If you have the means, purchase Energy Star products, and replace old "energy hogs." This would also be a good time to organize your fridge and pantry in order to reduce how much food you waste. The average American throws away 40 percent of the food they purchase. And while you're still in your kitchen, consider giving up disposable plates, napkins, and silverware!

In 2015, Bramble Cay melomys, rodents living in Australia, were the first animal known to be extinct due to human-caused global warming.

These may seem like a lot of changes, but each small step, along with sweeping changes by the government and environmental scientists, can keep our planet healthy for future generations.

At the end of *Under the Dome*, Julia Shumway must find a way to convince the "leatherheads" to let the remaining citizens of Chester's Mills live. She, finally, is able to convince one of the aliens that Julia and her fellow humans are beings with lives. That they *matter*, and deserve to live. This heart-to-heart works, and the dome is removed, allowing those trapped to breathe fresh air. One wonders if Earth, and its many creatures on the brink of extinction, could appeal to humans in the same way, if we would take our part in global warming more seriously?

SECTION FIVE
The 2010s

CHAPTER TWENTY-SIX

11/22/63

On November 22, 1963, President John F. Kennedy was killed in Dallas, Texas. The details of that day and the conspiracy theories surrounding it have fascinated people for decades. One of them is Stephen King. He was in high school at the time and recalled, "we got in the car and the radio was on . . . and the guy came on and said, 'the President is dead' and there was just total silence."[1] Ask anyone who was alive in 1963 where they were during the Kennedy assassination and they will probably be able to tell you.

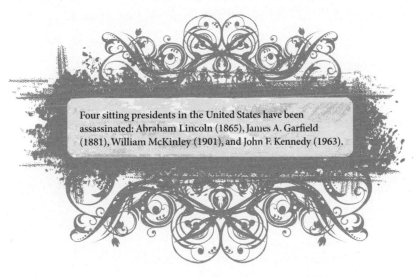

Four sitting presidents in the United States have been assassinated: Abraham Lincoln (1865), James A. Garfield (1881), William McKinley (1901), and John F. Kennedy (1963).

King got the idea for the book and said:

I'd like to tell a time-travel story where this guy finds a diner that connects to 1958 . . . you always go back to the same day. So, one day he goes back and just stays. Leaves his 2007 life behind. His goal? To get up to November 22, 1963, and stop Lee Harvey

Oswald. He does, and he's convinced he's just fixed the world. But when he goes back to '07, the world's a nuclear slag-heap. Not good to fool with Father Time. So, then he has to go back again and stop himself . . . only he's taken on a fatal dose of radiation, so it's a race against time.[2]

In the novel *11/22/63*, Jake Epping does indeed use a portal to travel back in time and try to prevent the Kennedy assassination. What is the science of time travel? To understand more about this complicated topic, we interviewed Dr. James Hedberg, a physicist and director of the City College of New York Planetarium.

Kelly: **"Time travel is featured in Stephen King's novel *11/22/63*. Now, while not all scientists believe time travel is possible, could you start by explaining Einstein's theory of time as a 'fourth dimension'? And what about his theory of relativity?"**

James Hedberg: "Time as a fourth dimension is an idea that pre-dated Einstein by many centuries, but, yes, is generally associated with his theory of relativity. Prior to his (and other contemporaries') work, time would have been considered a dimension that was independent of the other three spatial dimensions. The postulates of relativity however require that time and space no longer be independent dimensions but instead are joined, mathematically speaking, into one four-dimensional object known as spacetime. Phenomena such as time dilation and length contraction are a result of this joining."

Meg: **"NASA has a theory about travelers being able to go back in time via wormholes. Can you explain how this could be possible?"**

James Hedberg: "I'm guessing you might be referring to the breakthrough propulsion physics program, in which NASA engineers explored some modes of space travel somewhat more exotic than strapping astronauts to giant rockets. Terms like 'warp drive' and 'wormholes' were mentioned in some of the original project descriptions, but to my knowledge, these research topics are no longer being seriously explored at an official level by NASA. The basic idea of using a wormhole for time travel is predicated on having an 'arbitrarily advanced civilization' that could first construct a wormhole. This is well beyond the scope of our current technology. But, the mathematical framework from general relativity that has successfully been used to model much of what we see in the observable universe does appear to allow for the possibility of some type of folding of spacetime, the basis of wormholes, and also for time dilation effects to enable time travel, once the wormhole has been created."

Kelly: **"We read about astronomer Frank Tipler's idea of a cylinder to time travel. Are you familiar with his idea and do you think it could work?"**

James Hedberg: "I was not familiar with a Tipler cylinder before this. Would it work? I doubt it."

Meg: **"We've seen in some movies and TV shows that black holes or cosmic strings can be used to time travel. What is the scientific basis behind these theories?"**

James Hedberg: "The 'scientific' origin of all of these time travel sci-fi setups seems largely based on the observable fact that in the vicinity of very large (i.e., massive) objects like black holes, time does behave differently than our normal experience would suggest. Even near a rather small object like the Earth, engineers must account for relativity effects when programming the satellites in orbit. Near something much heavier, like a black hole, these effects will be more noticeable, and can easily serve as conceptual launching pads for more exotic (and much less verifiable) ideas like time travel. Though, as far as actually exploiting these mechanisms, I don't think there is much scientific basis."

Kelly: **"What is your favorite theory of time travel, either based in science or science fiction, and why?"**

James Hedberg: "I usually liked Star Trek's TNG time travel scenarios the best. They seemed to usually take care in dealing with the sticky issues of causality. In general, I love to imagine controlling our positions in time like we can our positions in space, but my science training doesn't make me think we'll ever have much luck making such things a reality. Most time travel scenarios in fiction seem to ignore the most obvious problem: that not only is Earth turning, but it is also moving around the sun, and the sun is moving around the galaxy, and the galaxy is moving through the universe, so were I to just blink back in time five minutes, the Earth would be somewhere completely different. But, yes, I do enjoy a good sci-fi time travel story."

Meg: **"Me too! Do you have a favorite Stephen King book or movie adaptation? What stuck with you about it?"**

James Hedberg: "I read a few as a kid, and they certainly did leave an impression. *Cujo* probably spent the most time bouncing around in my head growing up. I spent a lot of time in cars waiting for grown-ups to do something (back when it was okay to leave your kid alone in a car in a parking lot for hours . . .) and couldn't help but relate."

Kelly: "That would be much rarer to see nowadays! Thank you for explaining these complicated topics to us!"

There's a theory called the butterfly effect that could explain why things change each time Jake travels back in time. When Isaac Newton came up with the law of motion and universal gravitation in the 1600s, everything seemed predictable within science and the universe. This led to the belief that the future is set and we need to just wait for it to reveal itself. Determinism is the idea that the future is fixed while chaos theory purports that a small deviance could drastically change the outcome. The science of the butterfly effect is that tiny changes in big systems can have complex results.

For instance, in meteorology it's common practice to create several weather forecasts for predicting a hurricane's path because slight variables can create different results. A quote from Fichte in the 1800's *The Vocation of Man* explains the butterfly effect simply. "You could not remove a single grain of sand from its place without thereby . . . changing something throughout all parts of the immeasurable whole."

The Butterfly Effect theory purports that a very small change in initial conditions can create a significantly different outcome in weather or other occurrences.

There are numerous examples of the butterfly effect in action throughout history. The bombing of Nagasaki was originally supposed to take place in the city of Kuroko but was changed at the last minute due to cloudy weather conditions. An entirely different history would have played out if the initial plan had gone through. Adolf Hitler's future was also shaped by a seemingly small moment in time. He was an aspiring artist and applied to The Academy of Fine Arts in Vienna, Austria. He was rejected, twice, and it is said to be the catalyst for a shift in his worldview. The inciting incident that began World War I was the assassination of Archduke Franz Ferdinand. This event almost didn't happen, but, because of a botched first attempt, and a failure of communication with his driver, he was shot dead that day. The Chernobyl disaster is an example of how a small act saved millions. "After the initial explosion, three plant workers volunteered to turn off the underwater valves to prevent a second explosion. Had they failed to turn off the valve, half of Europe would have been destroyed and rendered uninhabitable for half a million years. Russia, Ukraine, and Kiev also would have become unfit for human habitation."[3]

Twenty-eight workers and firefighters succumbed to acute radiation poisoning during the first few months of the cleanup of Chernobyl, and dozens of others were badly sickened.[4]

What practical considerations would we need to think about if we were to time travel? In *11/22/63*, Al gives Jake the advice of picking up a pair of slacks instead of jeans and advises him to learn the lingo of the day. Both of these, era-accurate clothing and decade-specific language, would be imperative in order to not stick out in a crowd. Other considerations would be hairstyles, knowledge of the culture of the day, popular movies and books to reference, and overall manners for the specific year. Jake, traveling from 2011 back to 1958, would run into a host of variables that he would need to get used to. For example, a gallon of gasoline would have cost thirty cents in the late 1950s whereas a gallon in 2011 went for an average of $3.52. Fashion was vastly different in the 1950s and the culture of everyday conversation may have been challenging to adjust to. Through reading the book, we know Jake picks up on enough and doesn't out himself as a time traveler from the future.

Theorists haven't definitively agreed on what happened that fateful day in 1963, but that hasn't stopped everyday people from speculating. Stephen King traveled to Dallas, Texas, when conducting research for the book and came to some of his own conclusions. He believes Lee Harvey Oswald acted alone. When I (Kelly) was in high school I did my own research. I read the Warren Commission report, compared the notes of several experts, and pored over every piece of information I could get my hands on. My conclusion? I believed Lee Harvey Oswald may have been involved but didn't act alone. What science and evidence made me believe this? A few things.

First was the magic bullet theory. This theory posits that a single bullet caused all the wounds to Governor Connally and the nonfatal wounds to the president. The theory says that a three-centimeter-long copper-jacketed lead-core 6.5×52 mm Mannlicher–Carcano rifle bullet fired from the sixth floor of the Texas School Book Depository passed through President Kennedy's neck and went into Governor Connally's chest, went through his wrist, and embedded itself in the Governor's thigh. Assuming this is true, then this one bullet made its way through fifteen layers of clothing, seven layers of skin, and approximately fifteen inches of muscle tissue, struck a necktie knot, removed four inches of rib, and shattered a radius bone. The bullet was found on a gurney in the corridor at Parkland Memorial Hospital after the assassination in almost pristine condition. How could this be possible? A particular forensic technique used to match bullets found at crime scenes to bullets found in a suspect's possession, called comparative bullet lead analysis, was first used in this investigation.

In 2003, a panel of experts contradicted the FBI's analysis of the evidence at the time, and caused the bureau to stop using the technique altogether.[5] Why was this single bullet theory so important? If it was found that a second bullet had hit Governor Connally, it would have proven that there was more than one gunman. The Zapruder film shows the injuries caused on that day and the two men being wounded seconds apart wouldn't be consistent with a bolt-action rifle reloading in that period of time. The conclusion, therefore, would have to be that a single bullet caused all of the wounds or there was more than one shooter.

The second thing that I was fascinated with in studying the JFK assassination was the aforementioned Zapruder film, a silent 8mm color motion picture sequence shot by Abraham Zapruder. It's not the only footage from that day, but is the clearest example and was used endlessly in analysis. In a particular sequence, it appears that the president's head goes backwards from the impact of a bullet. This motion caused people to believe that perhaps there was another shooter who was in front of the motorcade. A 2018 study, which replicated the conditions and trajectory of bullets into melons, concluded that "the observed motions of President Kennedy in the film are physically consistent with a high-speed projectile impact from the rear of the motorcade, these resulting from an instantaneous forward impulse force, followed by delayed rearward recoil and neuromuscular forces."[6]

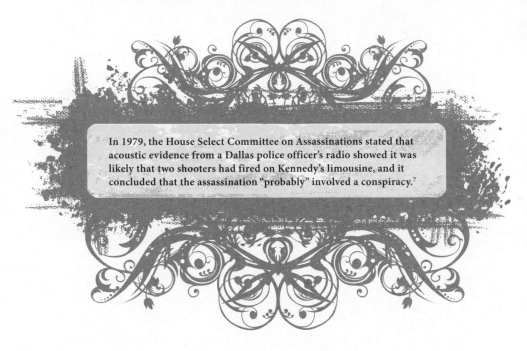

In 1979, the House Select Committee on Assassinations stated that acoustic evidence from a Dallas police officer's radio showed it was likely that two shooters had fired on Kennedy's limousine, and it concluded that the assassination "probably" involved a conspiracy.[7]

Last, I was intrigued by Lee Harvey Oswald's own claims that he was a "patsy." Could this be possible? Maybe it's the Fox Mulder conspiracy theorist in me, but I still believe it's possible he didn't act alone. There were several tests that were conducted after Oswald's arrest, one of them was a neutron activation analysis. This test can prove whether or not the person has fired a weapon recently. The test found no evidence of residues on Oswald's cheek, where the gun would have to be set in order to fire at the motorcade accurately. Seven marksmen carried out a test whereby they made the same shots as Oswald and after being tested, all of them had residues on their cheeks. The fact that Oswald didn't suggests he didn't fire a rifle that day. Another fact about Oswald is that he wasn't that great of a shooter. Having served in the United States Marines, his skills were tested. In 1956, his score of 212 was just two points above the sharpshooter requirement and well below the 220 needed for expert classification. Oswald's skills didn't improve as he scored just 191 in May of 1959. In subsequent years, sharpshooters have tried to replicate the lone gunman theory and failed.

Jake Epping prevents the Kennedy assassination in *11/22/63* but the world is in chaos after his return to 2011. He ultimately chooses to reenter the portal to undo his changes so that his beloved Sadie can live and the world will be set on its former path. We may not know what the future holds or how our decisions may impact the world but Stephen King makes one thing clear in this book: love is universal and transcends time and space.

CHAPTER TWENTY-SEVEN

Doctor Sleep

It would be easy to lose count of Stephen King's many revered novels and stories. So many have been developed into numerous iterations of TV series, movies, and even in the case of *The Shining*, an opera. He has developed book series like the Dark Tower and *Mr. Mercedes* trilogy. Most eagle-eyed constant readers can also find connections between many of his books, creating a complex web of mentions that correspond in different works. Yet, King is not known to write sequels for most of his most famous works. *The Shining* resonated with many readers, and thus was often brought up at readings, interviews, and fan events. One of the most common questions was whatever happened to Danny Torrance, the little boy who lost his father, and had the "shine," a unique power that allowed him to read thoughts and sense danger?

This question began to intrigue Stephen King. Danny would be grown up now. Would he be driven to addiction like his father? How would he handle his ability? Would it be more of a blessing or a curse? To his fans' great delight, King answered all these questions and more in the 2013 novel *Doctor Sleep*. Danny *had* unfortunately succumbed to alcoholism like his father Jack, yet after hitting rock bottom he found solace and purpose working in a hospice.

That is until he begins to communicate with a girl named Abra, who shares his "shining" as well as a telepathic connection with Danny. Things turn bad when the novel's villain, Rose the Hat, and her gang of vampires stalk Abra. Danny must become the hero once again, saving not only Abra, but every person who possesses the "shine." These vampires led by Rose the Hat are not the *Salem's Lot* variety. Instead of blood, they subsist on the fumes released by the fear of those with the power. This means they resort to torture and murder in order to stay young and healthy.

This obviously got us thinking about the extreme lengths people will go to appear younger or slow the process of aging. The term "fountain of youth" is often attached to fad diets, skin creams, or shakes full of spinach and vitamins. It is most popularly derived from the legend of Ponce De Leon and his exploration of what is now modern-day Florida. Juan Ponce De Leon was a Spanish conquistador who lived on the cusp of the sixteenth century. Stories have endured that Ponce De Leon was searching for the famed fountain of youth when he landed on the Eastern coast of Florida. Yet, historians today argue there is no proof of his motive.

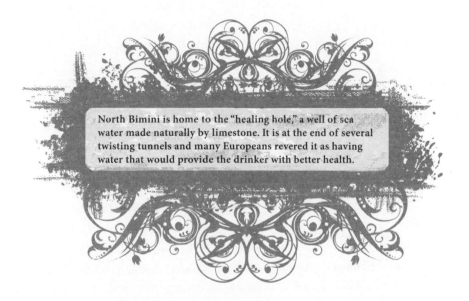

North Bimini is home to the "healing hole," a well of sea water made naturally by limestone. It is at the end of several twisting tunnels and many Europeans revered it as having water that would provide the drinker with better health.

Whether the Spanish conquistador was on the hunt for the fountain of youth, he was not the first to be linked to its existence.

Accounts of a river of youth were nothing new to men of the time. *The Alexander Romance*, a collection of stories about the mythical exploits of Alexander the Great, from the third century, has the conqueror looking for a river of immortality in India. And *The Voyage and Travels of Sir John Mandeville, Knight,* written in the fourteenth century, includes a description of the fountain of youth. When Caribbean natives spoke of an island called Bimini, where there was said to be a river of curative waters, some of the Spanish explorers believed them. Pietro Martire de

Anghiera, an Italian historian, discounted the story but wrote Pope Leo X in 1513 with his concerns that "all the people . . . think it to be true."[1]

Today, the site of the legend of Juan Ponce de Leon is a tourist attraction in St. Augustine, Florida. The Fountain of Youth Archaeological Park has no historical proof that Ponce De Leon was ever there, yet many visitors drink from the park's waters. Just in case it really does offer youthfulness, or perhaps even immortality!

Across the world from the lush forests of Florida was a castle in the snowcapped mountains of Hungary. Within its walls lived a Countess, desperate to regain her youthful beauty. While the legend of Ponce De Leon cannot be verified, the cruel actions of Hungarian Countess Elizabeth Bathory were witnessed by over three hundred, who were all willing to testify against her.

Living in a time of primitive understanding of science, the early seventeenth century, Bathory came to believe that she had the key to prolonging her youth. She simply needed the blood of young women. There is legend that she would bathe in it, as well as use the blood on her face like a salve or lotion. While this is not verified, her act of drinking blood seems to be proven. All in pursuit of maintaining her alabaster skin! This habit is why she has often been referred to as the "Real Countess Dracula." Her reign of torture began with torture and murder of female servants at her castle, eventually including women from nearby villages, as well as young teens sent by their parents to learn aristocratic manners from the Countess. "Bathory's torture included jamming pins and needles under the fingernails of her servant girls, and tying them down, smearing them with honey, and leaving them to be attacked by bees and ants. Although the Count participated in his wife's cruelties, he may have also restrained her impulses; when he died in the early 1600s, she became much worse. With the help of her former nurse, Ilona Joo, and local witch Dorotta Szentes, Bathory began abducting peasant girls to torture and kill. She often bit chunks of flesh from her victims, and one unfortunate girl was even forced to cook and eat her own flesh."[2]

These accounts may seem too cruel and outlandish to believe, but dozens of mutilated bodies were found on Countess Bathory's property. Despite her high connections, she was eventually discovered by local authorities. At trial she was convicted and although she murdered upwards of hundreds of women she was given the equivalent of house arrest. Elizabeth Bathory died three years later, at age fifty-four. There is no evidence to suggest her "trick" for eternal youth worked. Though, we have to guess the Countess would've most certainly joined Rose the Hat if it meant torture and less wrinkles!

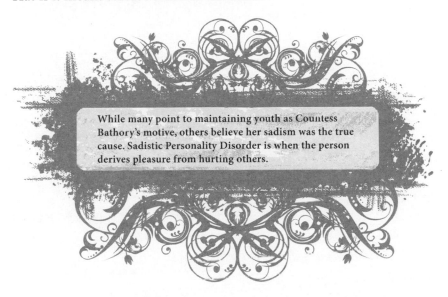

While many point to maintaining youth as Countess Bathory's motive, others believe her sadism was the true cause. Sadistic Personality Disorder is when the person derives pleasure from hurting others.

The pursuit of youth has undoubtedly caused people to go to extremes. Even today people are going under the surgeon's knife to smooth their bumps. But what if we could do something even more astonishing? The concept of astral projection is key in *Doctor Sleep*, as it is how Danny and Abra communicate. So, what is it exactly? If you go back to the idea of the soul, as discussed on page 104, one has to perceive our existence as being separate from our body. It is when this soul can leave and observe, or perhaps even interact with an environment outside of their body, that astral projection has occurred. Now, the most vital point to first understand is that there is no proof that astral projection, or out-of-body experiences, actually happen. Neurologists believe these feelings of dissociation, often discussed in terms of near-death experiences, are because of misfiring in the brain. While many others in numerous cultures disagree. At Gaia.com, a website devoted to yoga and meditation, there is a page explaining how to make yourself project astrally:

> There are dozens of methods to learn conscious OBE (out-of-body-experience) and astral projection. There are two approaches—one is to keep the mind awake while the body falls asleep. It's tricky—the mind wants to do what the body is doing. The goal is to take the body into deeper and deeper states of relaxation without drifting into unconsciousness. Yoga Nidra is one method. Once the body enters the sleep state, practitioners simply "roll" out of their physical form.[3]

They also share the "Rope Technique" developed by Robert Bruce, founder of the Astral Dynamics movement;

Step 1: Relax the physical body by visualizing each muscle.
Step 2: From your space of relaxation, enter a vibrational state; this should feel like an amplified version of a cell phone's vibration mode with pulsations coursing through the body.
Step 3: Imagine a rope hanging above you.
Step 4: Using the astral, or subtle, body, attempt to hold on to the rope with both hands. The physical body remains completely relaxed.
Step 5: Begin to climb the rope, hand over hand, all the while visualizing reaching the ceiling above you.
Step 6: Once you are aware of your full exit of the physical body, you are able to explore the astral plane.[4]

In the end, there is no proof in the scientific community that any of this is possible. In fact, another explanation for people believing they have actually left their body is the very real possibility of lucid dreaming. A lucid dream, experienced by nearly every human, is when the sleeper is aware they are in a dream.

Common elements of near-death experiences include a tunnel experience, floating above your own body, a feeling of peace, and bright white lights.

Thankfully, we have the fictional world of Stephen King, where astral projection can be a communication between two characters who need companionship, and eventually, help. It's King's imagination, a sort of meta-phorical astral projection, if you will, in which he creates stories and settings we can all visit. This brings his constant readers back for more. Even after nearly forty years, he lured us back to Danny Torrance, and the site of the burned Overlook Hotel.

Surveys suggest that between 8–20% of people claim to have had something like an out-of-body experience at some point in their lives.

CHAPTER TWENTY-EIGHT

Mr. Mercedes

As Stephen King has aged, his main characters have also become wiser and notably older. Such as in the case of Bill Hodges, a grizzled retired detective who has lost his zest for life. Faced with loneliness and boredom at the beginning of *Mr. Mercedes*, the divorced Hodges can't shake his final case before retirement. Somehow, a tech wizard with nefarious intentions used his knowledge to hack into a Mercedes, using the vehicle to kill innocent people at a job fair, including a young mother and infant. Haunted by the unsolved murders, old-school Hodges is faced with a technological world he doesn't understand. As the villain, Brady Hartsfield, mocks Hodges through untraceable means, the detective seeks help from younger and more tech fluent companions, Holly Gibney and Jerome Robinson.

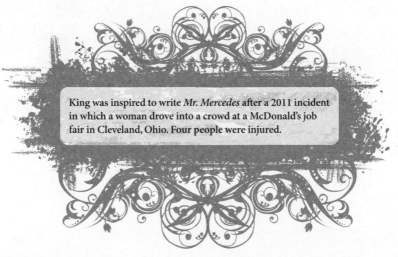

King was inspired to write *Mr. Mercedes* after a 2011 incident in which a woman drove into a crowd at a McDonald's job fair in Cleveland, Ohio. Four people were injured.

Mr. Mercedes is not the first time King has presented a world worsened by technology. In the end, it is the human spirit that prevails, as Holly finds the strength to save thousands of lives from Brady's bomb plot with the swift, simple swing of a weapon. In true King fashion, though, Brady is not stopped forever, and in the third book of the *Mr. Mercedes* trilogy, *End of Watch* (2016), comes back with a vengeance. Inspired to cause chaos, Brady utilizes a game app to send subliminal, suicidal messages to teenagers. Although this may seem far-fetched, subliminal persuasion is a real phenomenon, defined in the *Journal of Social Psychology* as "any word, image, or sound that is not perceived within the normal range of consciousness, but that makes an impression on the mind."[1] Many studies have been conducted on subliminal messages, particularly in relation to advertising. While there is no proof that extreme behavior such as suicide has occurred due to visual subliminal stimulation, there are examples of it worming into study participants' minds:

> For example, in the study by Naccache, Blandin & Dehaene (2002), subjects were required to decide whether a visually presented digit between the range of one and nine is greater or smaller than the target number five. Before the digits appeared, there was another digit which was presented subliminally. As a result, the decision speed was increased when the subliminally presented digit was congruent to the target stimuli by being greater or smaller than the digit five. The authors concluded that subliminal priming can activate a particular connection in the memory and make responses faster. However, the priming effect disappears if the length of time between priming and target stimulus is greater than one hundred milliseconds. Another very recent priming study was carried out by Friedman et al. (2005), in which males who were subliminally primed with words related to alcohol rated women as more attractive than when they were primed with words not related to alcohol. Priming words were presented for a few milliseconds on the screen before the attractiveness of a woman had to be evaluated in a subsequent picture. Words related to alcohol were, for example, "wine" or "beer," whereas words such as "coffee" or "milk" were not related to alcohol. Still, the priming effect could be observed only in subjects who preferred alcohol to stimulate their libido."[2]

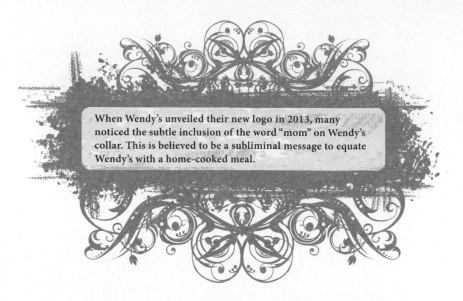

When Wendy's unveiled their new logo in 2013, many noticed the subtle inclusion of the word "mom" on Wendy's collar. This is believed to be a subliminal message to equate Wendy's with a home-cooked meal.

In order to further understand the world of modern technology, we talked to author and cyber expert R. J. Huneke:

Meg: **"Could you tell us about your background and how you got into cybersecurity?"**

R. J Huneke: "I have been making websites ever since I was first published as a columnist for *Newsday*'s impulse reviews section just over fifteen years ago. I had been building computers and modifying them since I was a kid. As a writer, I wanted to make myself as visible online as possible and so I took my fascination and used it as inspiration to build my own websites and increase my exposure while I was still in college. Cybersecurity is not my specialty trade; I have interviewed and consulted with some of the best in the business for my own books' research, and I would always defer to them in matters of cyber warfare. I have followed the scene closely over the last fifteen-plus years and I have helped a few companies out upon request when their sites were hacked, and I was able, through my knowledge of coding and research, to dispel some nasty adware. That said, I am not a certified cyber-security professional and those are the folks that you want to do penetration testing and security evaluations to keep you and your online presence safe."

Kelly: **"How have you brought your cyber know-how into your fiction writing? Was it an easy transition?"**

R. J Huneke: "For me it could not have been easier to fit cyber 'know-how' into my works of fiction, as the ever-evolving Cyberverse is truly remarkable and impactful. I am reading up on it every day, writing about gadgets, robotics, and technology frequently, and I am in awe of where we have come from and where we are going. That said, the world online scares the hell out of me. I am a big Ray Bradbury fan, and, yes, touching upon the incredible and dangerous Cyberverse could not have felt more natural to make a fictitious world that calls upon this dichotomy."

Meg: **"From Mary Shelley's *Frankenstein* to Stephen King's *Cell* and the *Mr. Mercedes* trilogy, horror has a long tradition of warning its readers about the pitfalls of science and technology. Why do you think that is?"**

R. J Huneke: "Stephen King is forthright in his criticism of technology and science with the near extinction of the human race due to a cell phone signal, in *Cell*, and the near mass suicide of children via subliminal advertising in a portable video game system in *End of Watch*, the last of the *Mr. Mercedes* trilogy. Though daunted by science-run-rampant at the outset of both tales, King has humanity prevail in the end of both stories. Throughout King's works there are frightening aspects of advancing tech present and deadly—how can we forget Blaine the Mono, a suicidal AI tired of life and jealous of two-legs that loves

Mary Shelley, in *Frankenstein*, warned her readers about the dangers of science and technology.

riddles and taunts Roland and his Ka-tet in *The Waste Lands* (1991) with a riddle competition that, alone, can save their lives? And throughout his works human beings can use the technology to their advantage or outright defeat the science with positive aspects that often come down to their being human."

Kelly: "It's amazing to me how Stephen King actually got me to hate a train!"

R. J Huneke: "He shows how an admittedly technologically ignorant person in the character of old Bill Hodges can overcome the most sinister of subliminal and telepathic abilities. While Stephen King calls out science and technology for the danger that it is, he does speak to the ability to use the positives of human nature, like creativity and tenacity, to overcome the calamities that can ensue once science is unleashed haphazardly."

Kelly: **"When you read a book or watch a movie, are you generally impressed by the media's portrayal of modern technology? Or are you often rolling your eyes at the inaccuracies?"**

R. J Huneke: "That's an interesting question. Most of the time when I read a book I do not find inaccuracies in portraying modern technology, and I think that is because when you read a really good book and the world-building is done right, your disbelief is suspended, and you are more invested in the characters' story than all of the nuances. I think about some of the greats, like Bradbury who wrote about reality TV, virtual reality without goggles, and giant screen televisions in the walls. He was very much ahead of his time in portraying modern science. Works like *The Expanse* (2015–), both on TV and the books (2011–), that rely on hard science and accepted theory and the use of gravity boots, for instance, make it so that there is enough real science and cutting-edge tech that you can get away with stretching some of the things out on the science-side a bit and not having the reader and/or viewer suspect something is amiss. I cannot tell you how often I am like, 'where is the explanation of how these spaceships have gravity,' because without boots holding them to the floor, these people should be floating all over the place. I will not name shows, movies, or books, but you know them. I recently read comments from a few people about an article on William Gibson's *Neuromancer* (1984) and how it was a good read, but dated, because it was written in the eighties."

Meg: "William Gibson wrote a very tech-based episode of our favorite show, *The X-Files!*"

R. J Huneke: "I have read the book multiple times and it remains one of the most innovative uses of language in world-building that I have ever come across and the sense of realism, while being surreal, is just so damn good. I love the book. It has never crossed my mind that Gibson, who coined the term 'cyberspace' and 'the matrix,' has technology in the book that has proven not to work out perfectly in the way modern technology has advanced in society. The noir thrill ride through the future that is *Neuromancer*, for me at least, sets up an entirely plausible and extremely visceral story that leaves a deep mark on the reader."

Meg: **"The character of Brady Hartsfield takes advantage of technology to wreak havoc, particularly in the King novel *End of Watch*, in which he employs an addictive game app to subliminally convince people to commit suicide. Tell us your thoughts on this modern form of murder. Have you witnessed subliminal messaging in tech? How realistic is *End of Watch*?"**

R. J Huneke: "I think *End of Watch* is very realistic, because subliminal messaging is proven to have stark effects on us, and if you are not buying into the unknown science of the brain with Brady's telekinetic abilities, we know more about space than we do the brain, and we know oh so little about space. You can still buy the story itself due to the magic woven in the world-building. King is a master storyteller who uses fully-fleshed characters of all walks of life and puts them on paper so that we can feel their breath on our faces when we get close to them. Before my first novel was picked up by a publisher I wrote a book about subliminal advertising and the research I did introduced me to things that blew my mind. *The Exorcist* (1973) employed CIA-tested techniques of subliminal sound messaging, for example, so that at tense moments sounds of swarms of bees or pigs being slaughtered are spliced into the soundtrack to further unnerve, disorientate, and scare the hell out of the viewer. McDonald's was caught some years back when someone noticed that when the program on the Food Network went to commercial break, there was a red blip on the screen. After hours of attempts a frame was freezed upon and the Mickey-D's bold red—the strongest of colors in advertising for getting attention—and the golden arches were shown with the message "I'm lovin' it" and that stood on the screen. The scariest thing about *End of Watch* is that it could really happen. I see a McDonald's commercial and, although I stopped eating most fast food years ago, I still crave that tasteless, salt-saturated poison and I have no idea why.

Meg: "Maybe that explains why my kids want McDonald's so often!"

Kelly: **"Mine, too! Can you give us any cybersecurity tips to avoid being victims of real life villains like Brady?"**
R. J Huneke: "I think the best thing anyone can do in an age of ransomware and identity theft is to educate themselves as much as possible. If you do not want to know a whole lot about cyber-security you can use a trustworthy cybersecurity professional for help, you can easily employ an antivirus, use a reputable password manager, and always create long, difficult passwords that are unique. Take full advantage of the free tools and information offered up by the nonprofit Electronic Frontier Foundation, EFF. org, where Privacy Badger can be installed into your browser of choice to block ads and ad trackers or the bots that take your information about what you see online and sell it to third party advertisers. I've been told by one of the best in the cybersecurity industry that anyone and anything can be hacked; it's just a matter of how much time, skill, and effort the hacker employs on the target—so you want to make it difficult and not use your birthday for a password."

Meg: **"What is your favorite Stephen King work and why?"**

R. J Huneke: "Now this question is just unfair—it's way too hard to choose a favorite work of King's. The short answer is *The Gunslinger*, the first book in the Dark Tower series. Though I think *Different Seasons* may be some of the finest fiction ever written by any author, and I love three of the four novellas in that book a great deal, and I love *Firestarter* so very much, too! *The Gunslinger* is unlike anything I have ever or will ever come across and it is truly special in myriad ways. I never thought I would find a book I loved as much as Tolkien's *The Lord of the Rings* which I re-read along with *The Hobbit* (1937) every year, for about twenty years now, and then I read *The Gunslinger*. 'The man in black fled across the desert, and the gunslinger followed.' That is the best first line in fiction. Period. The self-degradation of the soul alongside betrayal for one's cause, one's obsession is heart-wrenching, especially when it comes to Jake and Roland. And the mythology is so great, too.""

Speaking to someone with technological know-how who is also a Stephen King fan has been edifying. It's haunting to think that subliminal messaging is still used, and while it hasn't been proven to provoke mayhem, you never know!

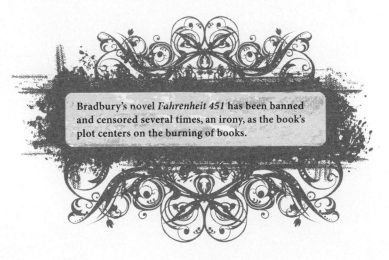

Bradbury's novel *Fahrenheit 451* has been banned and censored several times, an irony, as the book's plot centers on the burning of books.

Throughout the *Mr. Mercedes* trilogy, Bill Hodges feels lost in a world that has outgrown him. Technology alters his role as detective at a dizzying pace, often leaving him behind. Yet King is an expert at writing complex characters, and while Hodges at first seems incapable of change, he develops knowledge of technology, as well as acceptance of the paranormal. In the end, this fusion of old-school detective skills and the desire to understand technological science means the good guy wins again.

CHAPTER TWENTY-NINE

The Outsider

As with many of his stories, Stephen King started to think about the plot for his book, *The Outsider*, when he was taking part in an everyday activity. When his son Owen was twelve, King coached his little league baseball team. He recalled "nobody in a small town is more respected than someone who works with kids . . . and if something happens to that person where they find out they have a secret life that's not nice, nobody's more reviled and hated."[1] This inspired the character of Terry Maitland, a little league coach in Flint City, Oklahoma, to be arrested for the murder of a child. This unspeakable act leads the town to turn against this once beloved hero. But there's more to the story. Inspired by Edgar Allan Poe's *William Wilson* (1839), Stephen King wondered "what would a story be like if the evidence that somebody committed a horrible crime was ironclad. But if the evidence that the person had a perfect alibi, what if that was ironclad? You know, kind of an immovable object, an irresistible force."[2] Can a person be in two places at once? This is the central idea that begins the plot in the novel.

The 1839 story *William Wilson* by Edgar Allan Poe explores the idea of having a doppelganger.

There are two types of evidence found that prove Terry Maitland's whereabouts on the day of the murder. What type of evidence is more reliable: eyewitness testimony or DNA evidence? Without a doubt, DNA evidence has proven to be more reliable than eyewitness testimony. Why can't human memory necessarily be trusted? We all have short-term memory; similar to a Post-it Note, and long-term memory; akin to the hard drive of a computer. Eric Kandel, a neuroscientist at Columbia University in New York City, has shown that short-term memories involve relatively quick and simple chemical changes to the synapse that make it work more efficiently. He found that to build a memory that lasts hours, days, or years, neurons must manufacture new proteins to make the neurotransmitter traffic run more efficiently. Long-term memories must literally be built into the brain's synapses. Kandel and other neuroscientists have generally assumed that once a memory is constructed, it can't be undone.[3] This viewpoint is changing, though.

We tend to remember things through our own cultural filters and these memories can be skewed by a number of factors including bias, our mood, our physical health, and context. Karim Nader, a neuroscientist at McGill University in Montreal, believes that just the act of remembering can change the memories. In experiments, he has shown that our brains reconsolidate memories during the recall process and we can forget key aspects or even embellish them through conversations with others. Because of this, eyewitness testimony isn't as trustworthy as was once thought.

The advent of DNA analysis in the late 1980s revolutionized forensic science. It provides an unprecedented level of accuracy about the identity of actual perpetrators versus innocent people falsely accused of a crime. DNA testing led to the review of many settled cases in the United States and according to the Innocence Project, more than 360 people who had been convicted and sentenced to death since 1989 have been exonerated through DNA evidence. Of these, 71 percent had been convicted through eyewitness misidentification.[4]

DNA (deoxyribonucleic acid) is the human hereditary material that looks like a long molecule and it contains the information that organisms need to develop and reproduce. It was first discovered in

1869 by the biochemist Friedrich Miescher, who was studying the composition of lymphoid cells. It wasn't until the 1940s that genetic inheritance began to be researched and understood in regard to DNA. Human DNA is unique because it is made up of nearly three billion base pairs that are approximately 99 percent

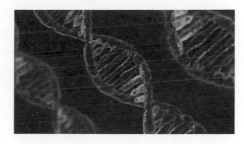

The human genome contains three billion base pairs of DNA.

the same in every person. Through advanced research, scientists believe the diagnosis and treatment of genetic diseases will continue to improve as well as the possibility of gene therapy. With personalized or precision medicine based on DNA, experts believe they will be able to pinpoint treatments for individuals much faster and more accurately.

Fingerprints are used to identify Terry Maitland and others in *The Outsider*. How accurate is fingerprint science? Fingerprints have been used throughout history as a means for identification but it wasn't until the nineteenth century that they were used to identify criminals. Since we all have unique ridges and patterns to our fingerprints, we are able to be identified if we touch a surface and leave a trace of sweat or other substances behind. Scientists are discovering that a lot more can be learned about a person through their fingerprints. In a new theory, scientists believe that fingerprints could leave a molecular signature behind. This could reveal aspects of an individual's lifestyle and environment, such as their job, their eating habits, or even their medical problems.[5] This is a fascinating new area of fingerprint science that will no doubt provide countless details about crime scenes and criminals in the future.

Another aspect of the detective work in *The Outsider* is comparing blood types found at the scene of a crime to the blood type of alleged perpetrators. How are blood types used in crime investigation? According to George Schiro, a forensic scientist at the Louisiana State Police Crime Laboratory, the most common applications of blood evidence are:

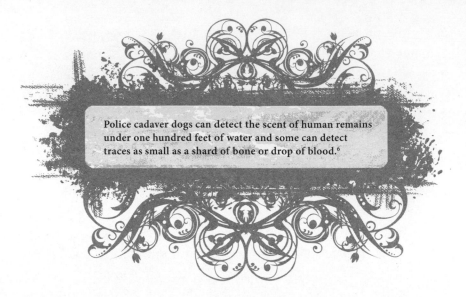

Police cadaver dogs can detect the scent of human remains under one hundred feet of water and some can detect traces as small as a shard of bone or drop of blood.[6]

Finding blood with the victim's genetic markers on the suspect, on something in the suspect's possession, or something associated with the suspect; finding blood with the suspect's genetic markers on the victim, on something in the victim's possession, or something associated with the victim; or investigative information determined from blood spatter and/or blood location.[7]

When it comes to determining blood types, there are four major groups determined by the presence or absence of two antigens, A and B, on the surface of red blood cells. There's also a protein called the Rh factor, which can be either present (+) or absent (–), creating the eight most common blood types: A+, A-, B+, B-, O+, O-, AB+, and AB-. Knowing a suspect's and victim's blood types can quickly determine who was present at a crime scene.

Holly Gibney plays a major role in the investigation in the book *The Outsider*. Her character description on the Stephen King Wiki page states "Holly suffers from OCD (Obsessive Compulsive Disorder), synesthesia, sensory processing disorder, and she's somewhere on the autism spectrum. Despite this, she's very observant, refreshingly unfiltered, and unaware of her innocence."[8] Obsessive compulsive disorder is a condition in which people have uncontrollable, recurring thoughts or behaviors.

It affects over two million people in the United States alone and can present itself with various symptoms. Some people who have OCD may have obsessive thoughts regarding germs, aggression, or having things be symmetrical. Others may have compulsions such as counting, repeatedly checking things, or excessively cleaning. Typically, OCD is treated with psychotherapy, medication, or a combination of the two.

What are the other conditions that Holly has? Synesthesia is a condition in which one sense is simultaneously perceived as if by one or more additional senses. Another form of synesthesia joins objects such as letters, shapes, numbers, or people's names with a sensory perception such as smell, color, or flavor. The word synesthesia comes from two Greek words, syn (together) and aisthesis (perception). Therefore, synesthesia literally means "joined perception."[9] Synesthesia can involve any of the senses but most commonly occurs for people to "see" colors for letters or numbers. Because there are multiple sense combinations, there are over sixty subtypes of synesthesia that could exist.

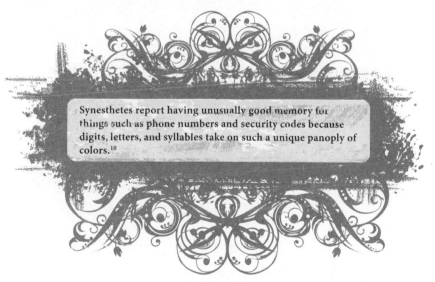

Synesthetes report having unusually good memory for things such as phone numbers and security codes because digits, letters, and syllables take on such a unique panoply of colors.[10]

Holly also has sensory processing disorder. People who have this condition are either very sensitive to things in their environment such as loud sounds, bright lights, or touch, or they could be unresponsive to these. Those with sensory processing disorder are often on the autism spectrum, as Holly is, but the condition itself has a wide range of possibilities. We haven't seen the last of Holly Gibney, as she appears in Stephen King's latest book, *If It Bleeds* (2020), which is a collection of four novellas.

The monster that is the villain in *The Outsider* is compared to several legends including El Cuco, Pumpkinhead, and others. What is the basis for these stories? The legend of El Cuco originated in Portugal. The oldest reference to the creature in writing dates back to 1274. The story is told in Portugal still today as well as in Spain and Latin America. Children are warned that if they misbehave, El Cuco will come to take them away and eat them. This legend is seen as the opposite of a guardian angel and is always present, watching and waiting to devour naughty children. The features of the creature are elusive because it is a shapeshifter. This makes for the perfect miscreant in *The Outsider* because it is able to commit its crimes by looking like anyone it wants to.

The Outsider ends with the creature being defeated and Terry Maitland's name being cleared. Even with the most unimaginable horror, Stephen King is able to give us a light at the end of the tunnel. Could such a creature exist in the world? According to King, "monsters are real, and ghosts are real too. They live inside us, and sometimes, they win." Thankfully for Ralph and Holly in this book, the monster didn't win.

CHAPTER THIRTY

The Institute

While Stephen King has masterfully tackled familiar frights: vampires, ghosts, and everything in between, in his sixty-first novel *The Institute*, the element of fear is rooted in something much more real: the government. As the author explains, the autumn of 2019 seemed like the perfect time to release *The Institute*. "All I can say is that I wrote it in the Trump era. I've felt more and more a sense that people who are weak, and people who are disenfranchised and people who aren't the standard, white American, are being marginalized," King said in a recent *New York Times* interview. "And at some point in the course of working on the book, Trump actually started to lock kids up . . . that was creepy to me because it was really like what I was writing about."[1]

While King admits there is an obvious allegory one could draw from Trump's border policy to the children trapped in his novel, he makes certain to point out that this was not his original intention, but rather a fortuitous accident. He further asserts in the *Guardian*, "children are imprisoned and enslaved all over the world. Hopefully, people who read *The Institute* will find a resonant chord with this administration's cruel and racial policies."[2]

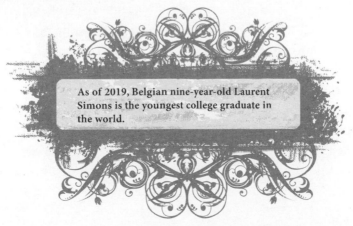

As of 2019, Belgian nine-year-old Laurent Simons is the youngest college graduate in the world.

The novel centers on Luke Ellis, a highly precocious preteen who is on the cusp of college at only twelve years old. One night, Luke's life is forever changed when strangers, who we learn are employed by the mysterious institute, murder his parents and kidnap him. Once awake in a strange, school-like building, Luke meets a group of children who have met the same terrible fate. In fact, it is not Luke's unique intelligence that is the cause of his capture, but rather his tendency to break objects when angry. With his mind, of course.

While Luke doesn't have the impressive powers of Carrie White or even *Firestarter*'s (1980) Charlie McGee, he is of great interest to the researchers of the Institute. There is already a file of information on Luke's life and abilities, including the fact that he is TK, shorthand for telekinetic. Forced to undergo painful shots, humiliating medical tests, and most brutal of all, no contact with the outside world, the kids of the Institute quickly become friends. They share in one simple goal, to escape the fences beyond the industrial building and regain their freedom.

Luke's friends include Avery, petite and scared from his new circumstances. Avery's telepathic power, or, as he's known in the novel, TP pos, is more potent than Luke's. Friend Kalisha, too, holds a telepathic power, rooted in a deep empathy, that will aid in their escape. But, before Luke and friends can work against the people at the Institute, the reader is introduced to the "Back Half," a building beyond, where kids disappear to, and never return from. While the "Front Half'" is not pleasant, the "Back Half" is where the true Stephen King horror blossoms. There, kids as young as six, both TP and TK, are used as weapons against American enemies. Their minds are harnessed into a supernatural energy, which is then directed to murder. Eventually, this breaks down the children's psyche. They lose their personality, their mind. They are husks of their former selves. It is the worst kind of zombie, bred by the government and then left to suffer after their task is done.

Politics aside, the plot of *The Institute* mirrors widely known experiments conducted by the CIA decades ago. While the real American government certainly would not admit to any sort of similar doings, they *have* released information on a controversial top-secret project entitled MK-Ultra. According to History.com, "MK-Ultra was a top-secret CIA project in which the agency conducted hundreds of clandestine experiments—sometimes on unwitting US citizens—to assess the potential use of LSD and other drugs for mind control, information gathering, and psychological torture. Project MK-Ultra lasted from 1953 until about 1973."[3]

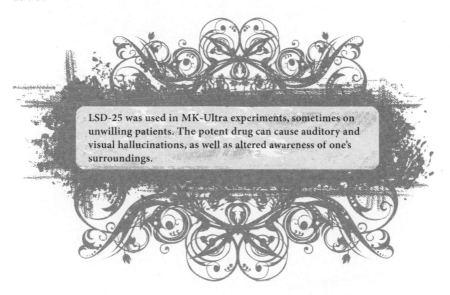

LSD-25 was used in MK-Ultra experiments, sometimes on unwilling patients. The potent drug can cause auditory and visual hallucinations, as well as altered awareness of one's surroundings.

It would be reassuring to believe that studies of telekinesis exist in only fiction, yet in 1975, the documents on MK-Ultra were released, outlining twenty years of tax-funded research.

Most upsetting is that in the decades following MK-Ultra experiments, it has become clear that the participants were inextricably harmed. Nineteen-year-old Phyllis Goldberg, barely older than the kids in *The Institute*, was a bright nursing student when she sought help for depression. She was soon pulled into the research of Dr. Donald Ewen Cameron, the president of both the American and Canadian

Donald Ewen Cameron was an MK-Ultra contributor.

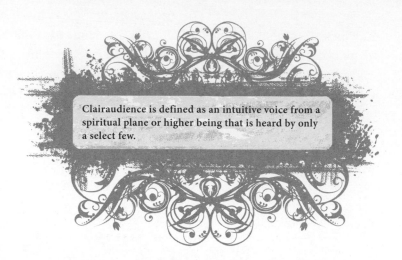

Clairaudience is defined as an intuitive voice from a spiritual plane or higher being that is heard by only a select few.

Psychiatric Associations, who unbeknownst to his patients, was an MK-Ultra contributor. The CIA had begun to pay him to further his work on psychic driving, in which he believed he could manipulate the mind to forget memories and recircuit new pathways. The CIA hoped this potential breakthrough would assist in placing pressure on spies.

In Dr. Cameron's pursuit, Phyllis and fellow patients suffered a barrage of drugging, shock therapy, induced comas, and more indignities. This led to horrific consequences. They "would suffer extreme personality changes, incontinence, amnesia, and in many cases, revert them to a state of child-like dependency."[4] As Phyllis's niece, Marlene Levenson, recounted for CTV, her aunt Phyllis was never the same. "When she would be with us, on weekends and so on, she didn't communicate. She laughed for no reason. Her gait was very different. She couldn't dress herself—she couldn't do anything for herself." This led to Phyllis spending the last twenty years of her life in a vegetative state. Not unsurprisingly, Phyllis's story is not unique, and has led her family and other victims to seek justice with a class action lawsuit.

It is important to note that in the documents given to the public, there is no proof that the life-altering experiments conducted by Dr. Cameron on his subjects led to success for spy interrogations in the CIA.

MK-Ultra is not the American government's last attempt to capture the power of psychics. Uri Gellar, a popular television purveyor of ESP ability in the 1970s, caught the attention of the CIA. He, and others like him, were asked to be a part of a top-secret project in which they would be used, not unlike the children in *The Institute*, to find defectors. One notable difference, as far as what is known, is they were all adults who were asked, *not forced*, to cooperate. Annie Jacobson, author of *Phenomena: The Secret History of the U.S. Government's Investigations into Extrasensory Perception and Psychokinesis* (2017), explained the merit the government sought in people with supposed ESP. "This is where it got very interesting, because scientists would consider, 'wait a minute, maybe we can read the minds of other government officials; maybe we can see inside a nuclear facility in Russia.'" As she continued in her interview with Erin Moriarity on *CBS Sunday Morning*, Jacobson said that despite the government's open mind, "There's instances of unusual situations, but there is no proof. It does not pass scientific muster."

Scientist Dean Radin, author of *Real Magic: Ancient Wisdom, Modern Science, and a Guide to the Secret Power of the Universe* (2018), disagreed with Jacobson in the piece for CBS, stating "What we're talking about is something like a talent, similar to musical talent or sports talent. So, there will be some people who are at the Olympic level; most of us aren't there."[5]

To get to the bottom of this fascinating topic, we spoke to clairvoyant Bonnie Macleod about her experiences as an empath.

Meg: **"How would you classify your psychic/medium ability? Is it something you have control over, or more of a sudden sense or feeling?"**

Bonnie Macleod: "I consider myself a clairvoyant empath more than anything. I also have clairaudient experiences though they are rare for me. I always describe my experience as a 'knowing.' I get flashes of items, people, and places in my mind that are accompanied by physical experiences (pain, joy, sadness, etc.) and a general 'knowing,' much like when a person talks to themselves in their own head (think-speak). I do have control over things should they become overwhelming (physical pain, or graphic images, or just plain inconvenience). I will say that I cannot do it at will. I can't force anything to cooperate or happen and won't pretend to either. I don't guess at things, if I'm getting nothing I just get nothing. It doesn't happen often, but it does happen. I don't put on a show. I'm not a trick pony."

Kelly: **"Can you describe the first time you realized you had a unique ability?"**

Bonnie Macleod: "I was twelve or thirteen years old and was lying in my bed, wide awake, when suddenly a face appeared in my face almost nose-to-nose, and smiled at me and slowly faded away. I did not feel scared but rather peaceful and calm. I got out of bed and opened my bedroom door and at the same time my mom opened hers, and we looked at each other for a moment and then she said, "You saw him too, didn't you?" And I said, 'yes, who was that?' And she said, 'that was your grandpa.' My grandpa passed away when I was only a year old. That was the first time that I remember the experiences starting, although I've had an irrational fear of the dark since childhood so it's possible I had earlier experiences."

Meg: **"Why have you chosen not to monetize this trait like many have?"**

Bonnie Macleod: "After my first experience, many people have appeared to me. I've also had many experiences telling people about their deceased family members who I never met and shouldn't have known about. It happened so many times and so often that when I reached my twenties, I decided to try to accept that part of myself instead of feeling like a freak that people would judge. Although I am not religious now, I had not been to church for a good number of years, but in order to accept this gift completely, I found myself needing to be sure it wasn't something that made me 'evil' or 'corrupted' in any way. I picked up the phone and called the preacher of the church I had attended in my childhood to ask him if he thought this gift made me evil in the church's eyes. He said that in his opinion one cannot be a true Christian and not believe that such gifts are given by God to some of his children and that it's human corruption and greed that have led some people to see these gifts as evil. He went on to say that he knew I had these gifts because of some things I had said to him as a child. He told me to use them to help people, as that's how they are intended to be used. I then decided that help is just that, help. Not profiting from pain, not profiting from need. So, I knew I could not profit from this gift."

Kelly: **"What reactions do you get when people learn about your gift?"**

Bonnie Macleod: "I've gotten mixed reactions, as one might expect. People doubt what they can't comprehend. I will say though, for those I have helped, they believe and know without doubt. I don't ask questions, I just simply tell them what comes to me. My sister calls it my 'scary stuff!' My mother isn't surprised, it runs in the women in my family (not all of them). My dad had an experience with me after I had a surgery and was starting to wake up that brought him to tears and he never doubted it was true. I was not aware of what I was saying and told him of talking with his deceased brother, giving him details and a message that I shouldn't have known. A lot of people doubt the existence of life after death and I accept that. My response is that I know it exists because I experience it. No one can tell me otherwise!

Meg: **"There are many types of telepathic and telekinetic abilities portrayed in media, Stephen King's work included. In *Carrie*, Carrie White can move objects, Charlie McGee spreads fire in *Firestarter*, and many of his characters have communed with the dead from *The Shining* to *Pet Sematary*. How do you feel the media, Stephen King, and others, have treated those like yourself?"**

Bonnie Macleod: "I think that King demonstrates a fairly good balance of both the good and bad things of being a medium. I think one of the best examples of my experiences is in *Pet Sematary*, the example of the main character's dead patient coming back to give him a message. He is portrayed as scary and devious but is actually trying to help, not hurt the family. Perhaps the very best example of my experiences and abilities is in *The Dead Zone* (1979). That book is a very good representation of how things actually occur for me. The 'flashes' I have are much like in old movies when they would flash 3, 2, 1 on the screen before the movie starts. Images flash like that to me at times and at times it's the actual person I see."

Kelly: **"In *The Institute*, children with impressive psychic abilities are captured by the government in order to be tested and used as weapons. Has there been a time you felt someone wanted to exploit your trait for their personal gain?"**

Bonnie Macleod: "Honestly, no. I've never had someone try to take advantage of my gifts. Most of the time I call them or visit them after an experience, of course it's mostly friends and family."

Meg: **"What would you like people to know about the reality of living with your gift? What are some false ideas people seem to have?"**

Bonnie Macleod: "I would like people to understand that not all mediums are 'guessing' or trying to profit from them. I also want people to know that at times it can feel like a burden when people say hurtful things. It's not something I asked for, and it took a long time for me to accept it and embrace it as a part of me. It's also not a parlor trick, nor is it on demand. I refuse to guess or manipulate just to make someone feel good. Sometimes there just isn't any message, it's rare but it happens."

Kelly: **"Lastly, what is your favorite work by Stephen King and why?"**

Bonnie Macleod: "I love both *11/22/63* and *The Long Walk* (1979) equally. They both leave the reader questioning how they would deal with those situations. They make you question exactly how selfless you could be, how strong you could be, and make you realize how easy it is to fall in love with a character in a book and embrace them as family."

Thank you to Bonnie for insight into such a fascinating world.

The children of *The Institute* don't have a choice about their abilities, or, like the victims of the MK-Ultra tests, whether they will be poked and prodded. Yet, Stephen King gives the fictional kids in his novel the ability to fight against what often feels impossible to rise against, a well-oiled and powerful government.

Conclusion

Thank you, fellow constant readers, for joining us on this journey into the heart of Stephen King and the complex, horrific worlds he has created. Truth is often stranger than fiction, and for the Master of Horror, he fused both, dropping memorable characters into plots as vast as killer clowns to time travel in the era of the JFK assassination. As the author himself said, "fiction is the truth inside the lie." It is this intersection of both his believable, empathetic characters and his fictional, often paranormal horror that makes him the true master.

Acknowledgments

Thank you to Nicole Mele and everyone at Skyhorse!

Thank you to our parents who introduced us to Stephen King through his books and movies.

Thank you to everyone who took the time to be interviewed including Les, Luke, Andrew, Annette, James, Kara Lee, Richard, R. J, Amanda, Sara, Samantha, and Bonnie.

Thank you to our families for all of their love and support. And to the Rewinders, we'll see you in the horror section.

About the Authors

Kelly Florence (left) and Meg Hafdahl

Kelly Florence is a communication instructor at Lake Superior College in Duluth, Minnesota, and is the creator and cohost of the *Horror Rewind* podcast as well as the producer and host of the podcast *Be A Better Communicator*. She received her BA in theater at the University of Minnesota–Duluth and earned her MA in communicating arts at the University of Wisconsin–Superior. Kelly is the coauthor of *The Science of Monsters* and *The Science of Women in Horror*, also from Skyhorse Publishing.

Horror and suspense author Meg Hafdahl is the creator of numerous stories and books. Her fiction has appeared in anthologies such as *Eve's Requiem: Tales of Women, Mystery, and Horror* and *Eclectically Criminal*. Her work has been produced for audio by *The Wicked Library* and *The Lift,* and she is the author of two popular short story collections including *Twisted Reveries: Thirteen Tales of the Macabre*. Meg is also the author of the two novels *Daughters of Darkness* and *Her Dark Inheritance,* called "an intricate tale of betrayal, murder, and small-town intrigue" by *Horror Addicts* and "every bit as page turning as any King novel" by *RW* magazine. Meg, also the cohost of the podcast *Horror Rewind* and coauthor of *The Science of Monsters* and *The Science of Women in Horror,* lives in the snowy bluffs of Minnesota.

Endnotes

Chapter One: Carrie

1. Romano, Aja. (October 10, 2018) "Stephen King: A Guide to His Horror, His History, and His Legacy." *Vox.*
2. Abrams, Bryan. (October 18, 2013) "The Carrie Phenomenon: A Brief History of Telekinesis." *Motion Pictures.org.*
3. King, Stephen. (2000) *On Writing: A Memoir of the Craft.* New York: Scribner.
4. Lawson, Carol. (September 23, 1979) "Behind the Best Sellers: Stephen King." *The New York Times.*
5. (2020) www.stephenking.com
6. Creed, Barbara. (1993) "Horror and the Monstrous." Routledge: London.
7. Fransson, Rebecka. (2015) "Bloody Horror! The Symbolic Meaning of Blood in Stephen King's Carrie." Orebro University.
8. Adams, Zoe. (August 31, 2015) "Diseasing the Female." *Eidolon.*
9. Dua, P. (March 2014) "Orbital Vicarious Menstruation." *NCBI.*
10. Graziottin, Alessandra. (August 1, 2016) "Perimenstrual Asthma: From Pathophysiology to Treatment Strategies." *Multidisciplinary Respiratory Medicine.*
11. Zaslavsky, Claudia. (January 1992) "Women as the First Mathematicians." *International Study Group of Ethnomathematics Newsletter.*
12. Ilias, I. (2013) "Do Lunar Phases Influence Menstruation? A Year-Long Retrospective Study." *Endocrine Regulations.*
13. Howarth, Jan. et al. (November 28, 2008) "Religion, Beliefs and Parenting Practices." *Joseph Rowntree Foundation.*
14. Dawkins, Richard. (2006) *The God Delusion.* Bantam Books: United Kingdom.
15. VanderWeele, Tyler J. (September 18, 2018) "Religious Upbringing and Adolescence." *Institute For Family Studies.*
16. Tashjian, Sarah. (May 16, 2018) "Parenting Styles and Child Behavior." *Psychology in Action.*
17. Swearer, Susan M. (May 2015) "Understanding the Psychology of Bullying." *American Psychologist.*
18. Zhang, Peng. (May 20, 2016) "Social Anxiety, Stress Type, and Conformity Among Adolescents." *Frontiers in Psychology.*
19. Walton, Alice G. (February 21, 2013) "The Psychological Effects of Bullying Last Well Into Adulthood, Study Finds." *Forbes.*
20. King, Stephen. (1974) *Carrie.* Doubleday: New York.

Chapter Two: The Shining

1. (July 2, 2010) "Stephen King: The 'Craft' of Writing Horror Stories." *NPR*.
2. Beahm, George Andrews. (1998) *Stephen King: America's Best-Loved Boogeyman*. McMeel Press.
3. Derin, Jacob. (March 18, 2018) "The Shining: The Hellish World of the Tyrannical Patriarch." *Medium*.

Chapter Three: Salem's Lot

1. Konstantin, Phil. (July 1987) "An Interview with Stephen King." *Highway Patrolman Magazine*.
2. Tucker, Abigail (October 2012) "The Great New England Vampire Panic." *Smithsonian*.
3. Dolin, [edited by] Gerald L. Mandell, John E. Bennett, Raphael. (2010) *Mandell, Douglas, and Bennett's principles and practice of infectious diseases* (7th ed.). Philadelphia, PA: Churchill Livingstone/Elsevier.
4. Austen, Barbara. (December 3, 2014) "Consumption." *Connecticut Historical Society*.
5. Tucker, Abigail. (October 2012) "The Great New England Vampire Panic." *Smithsonian*.

Chapter Four: Rage

1. (2020) *Stephen King.com*
2. Katisyanis, Antonis. (July 15, 2018) "Historical Examination of United States Intentional Mass School Shootings in the 20th and 21st Centuries: Implications for Students, Schools, and Society." *Journal of Child & Family Studies*.
3. Warnick, Bryan R. (January 1, 2020) "Protecting Students from Gun Violence: Does "target hardening" do more harm than good?" *Education Digest*.
4. King, Stephen. (January 25, 2013) "Guns." Philtrum Press.
5. Ferguson, Christopher, (January 1, 2008) "The School Shooting/Violent Video Game Link: Causal Relationship or Moral Panic?" *Journal of Investigative Psychology & Offender Profiling*.
6. Anderson, Craig A. (2003) "Exposure to Violent Media: The Effects of Songs With Violent Lyrics on Aggressive Thoughts and Feelings." *Journal of Personality and Social Psychology*.

Chapter Five: The Stand

1. (2020) "The Stand: Inspiration." *StephenKing,com*.
2. Woolf, Jim. (January 1, 1998) "Army: Nerve Agent Near Dead Utah Sheep in '68; Feds Admit Nerve Agent Near Sheep." *The Salt Lake Tribune*.
3. Hooker, Edmund. (January 10, 2019) "Biological Warfare." *EMedicine Health.com*.

4. Geggel, Laura. (April 07, 2017) "7 Facts About the Deadly Nerve Agent Sarin." *Live Science.*

5. (June 11, 2019) "Smallpox Fast Facts." *CNN.*

6. (February 5, 2020) "The Spanish Flu." *History.com.*

7. Gunderman, Ricard. (September 14, 2018) "The "Greatest Pandemic in History" Was 100 Years Ago – But Many of Us Still Get the Basic Facts Wrong." *Healthline.com.*

8. Brazier, Yvette. (May 22, 2018) "Everything You Need to Know About Pandemics." *Medical News Today.*

Chapter Six: Nightshift

1. (November 11, 2004) "Images of Desire: Brain Regions Activated By Food Craving Overlap With Areas Implicated In Drug Craving." *Science Daily.*

2. Sachan, Dinsa. (September 21, 2019) "Why Do We Crave?" *Brain World Magazine.*

3. Stetka, Bret. (April 16, 2016) "The Human Body's Complicated Relationship With Fungi." *NPR.*

4. Stetka, Bret. (April 16, 2016) "The Human Body's Complicated Relationship With Fungi." *NPR.*

5. Striepe, Becky. (June 15, 2011) "10 Organizations That Want to Help You Quit Smoking." *How Stuff Works.com.*

6. Andritsou, Martha. (2016) "Success Rates are Correlated Mainly to Completion of a Smoking Cessation Program." *European Respiratory Journal.*

7. (2020) "Quitting Smoking." *Centers for Disease Control and Prevention.gov.*

Chapter Seven: Cujo

1. (2020) "Cujo: Inspiration." *StephenKing.com.*

2. King, Stephen. (2000) *On Writing.* Scribner: New York,

3. Stone, Zara. (January 27, 2016) "The Pet Funeral Industry Makes 100 Million in Profit." *The Hustle.*

4. Turner, Josie F. (February 23, 2020) "Is it Legal to Own a Wolf-Dog? Find out Everything about this Hybrid." *Animalwised.com.*

5. Bula-Rudas, Fernando J.; Olcott, Jessica L. (October 1, 2018) "Human and Animal Bites." *Pediatrics in Review.*

6. Mullaney, Julia. (July 11, 2018) "This is What Really Happens to Your Body in a Hot Car." *Cheat Sheet.*

7. (October 17, 2018) "Pet Vaccinations: Facts and Myths." *Pets Best.com.*

8. (November 23, 2016) "Louis Pasteur and the Development of the Attenuated Vaccine." *VBI Vaccines.*

Chapter Eight: Pet Sematary

1. (2020) "Pet Sematary: Inspiration." *Stephen King.com.*

2.   MacKinnon, Eli. (February 8, 2012) "Is It Possible to Reanimate the Dead?" *Live Science.*

3.   Swancer, Brent. (March 2, 2015) "The Dark Quest to Reanimate the Dead." *Mysterious Universe.*

4.   (2020) "Spinal Meningitis." *Cedars Sinai.edu.*

5.   (2020) "Facts About Meningitis." *Confederation of Meningitis Organisations.*

6.   Freneau, Philip, (1787) "The Indian Burying Ground." *Poetry Foundation. org.*

7.   Stone, Zara. (January 27, 2016) "The Pet Funeral Industry Makes 100 Million in Profit." *The Hustle.*

8.   King, Stephen. (1983) *Pet Sematary.* Doubleday: New York.

Chapter Nine: Thinner

1.   (February 2020) "Thinner Inspiration." *Stephen King.com.*

2.   Shahbandeh, M. (January 14, 2019) "U.S. Diets and Weight Loss - Statistics & Facts." *Statista.*

3.   Lallanilla, Marc. (August 6, 2013) "Real or Not? 6 Famous Historical Curses." *Live Science.*

4.   (2020) "Pedestrian Safety." *CDC.gov.*

5.   LeBeau, Phil. (February 28, 2019) "Pedestrian Deaths Hit 28-Year High." *CNBC.*

6.   (February 2020) "Adult Obesity Facts." *Center for Disease Control and Prevention.*

7.   Sampson, Stacy. (September 28, 2018) "Parasitic Worms in Humans." *Healthline.*

8.   (November 20, 2019) "Tapeworms in Humans: Causes, Symptoms, and Treatments." *Carygastro.com.*

9.   (2020) "Eczema Statistics." *MG217.com.*

Chapter Ten: It

1.   Kozlowski, David. (August 31, 2017) "Stephen King Talks About The Horror of Pennywise and IT." *LRM Online.*

2.   Wilding, Mark. (December 29, 2019) "Having a Laugh: Is this the end for Clowning?" *The Guardian.*

3.   (2020) "Commedia dell'Arte Characters." *Italy Mask.co.*

4.   Meiri, Noam et al. (February 1, 2017) "Fear of clowns in hospitalized children: prospective experience." *European Journal of Pediatrics.*

5.   Freud, S. (1955) *The Uncanny in Art and Literature* (2nd Edition). London: Penguin Books.

6.   Hermann Ph.D., Henry R. (2017) "Alternate Human Behavior in Dominance and Aggression in Humans and Other Animals." *Science Direct.com.*

7.   Reyes, Mike. (September 8, 2017) "Master of horror Stephen King revealed his personal feelings on clowns." *CinemaBlend.*

Chapter Eleven: The Drawing of the Three

1. Browning, Robert. (1895) "Childe Roland to the Dark Tower Came." *A Victorian Anthology.* Cambridge: Riverside Press.
2. Armenti, Peter. (August 18, 2017) "How Did Stephen King to the Dark Tower Come? Through Robert Browning's 'Childe Roland.'" *Library of Congress.gov.*
3. (2020) "Dissociative Disorders." *National Alliance on Mental Health.*
4. (July 2018) "Drugs, Brains, and Behavior: The Science of Addiction." *National Institute on Drug Abuse.*
5. Thoma, Scot. (February 3, 2020) "Alcohol and Drug Abuse Statistics." *American Addiction Centers.org.*
6. (2018) "Heroin." *The National Institute on Drug Abuse.*
7. Brooks, Megan. (June 17, 2015) "Drug Overdose Now Leading Cause of Injury-Related Deaths." *Medscape.*
8. Guth, Alan. (March 18, 2014) "Alan Guth Explains Inflation Theory." *World Science Festival.*
9. Vyas, Kashyap. (August 27, 2018) "String Theory Explained: A Brief Overview for Starters." *Interesting Engineering.*
10. King, Stephen. (1982) *The Dark Tower: The Gunslinger.* Grant: Hampton Falls.

Chapter Twelve: Misery

1. Wiel van der, Mark. (1350) "A Broken Leg in the Year 1350: Treatment and Prognosis."
2. Vic. (October 8, 2010) "Setting a Broken Bone: 19th century medical treatment was not for sissies." *Jane Austen's World.*
3. Gallimore, Joe. (August 4, 2017) "How Stephen King's 'Misery' Predicted Modern Fandom." *Bloody Disgusting.*
4. Guarniero, Bevilacqua. (June 2017) "The schizophrenia stigma and mass media: a search for news published by wide circulation media in Brazil." *International Review of Psychiatry.*
5. Chan, Sherry Kit. (January 2019) "The effect of media reporting of a homicide committed by a patient with schizophrenia on the public stigma and knowledge of psychosis among the general population of Chinese Hong Kong." *Social Psychiatry & Psychiatric Epidemiology.*
6. Chan, Sherry Kit. (January 2019) "The effect of media reporting of a homicide committed by a patient with schizophrenia on the public stigma and knowledge of psychosis among the general population of Chinese Hong Kong." *Social Psychiatry & Psychiatric Epidemiology.*

Chapter Thirteen: The Tommyknockers

1. Greene, Andy. (October 31, 2014) "Stephen King: The Rolling Stone Interview." *Rolling Stone.*

2.   Moore, Kate. (February 1, 2019) The Radium Girls: The Dark Story of America's Shining Women. Sourcebooks: Chicago.
3.   Wallace-Wells, David. Et al. (March 20, 2018) "13 Reasons to Believe Aliens are Real." *NY Mag.com*.
4.   (April 2020) "UFO Story." *Roswell-NM.gov*.
5.   (November 9, 2009) "Roswell." *History.com*.

Chapter Fourteen: The Dark Half

1.   Smythe, James. (October 21, 2013) "Re-reading Stephen King." *The Guardian*.
2.   Brown, Steve. (2019) "Bachman Exposed." *Lilja's Library*.
3.   Bogdan, Robert. (1990) *Freak Show: Presenting Human Oddities for Amusement and Profit*. University of Chicago Press.
4.   Strickland, Ashley. (March 21, 2017) "Risky Surgery Separates 10-month-old From Parasitic Twin." *CNN*.
5.   Galistan, Diane. (September 2, 2019) "Teenager endured her parasitic twin for five years before mass of flesh and bones was discovered." *International Business Times*.

Chapter Fifteen: Needful Things

1.   (2020) "Soul." *Merriam Webster*.
2.   Morse, Donald R. (July 1, 2005) "Can Science Prove the Soul, the Afterlife and God?" *Journal of Religion & Psychical Research*.
3.   DeeKay, Dheeraj. (April 24, 2016) "Conversations: On Existence of Soul and Two 'Scientific' Experiments That Prove One!" *Extra News Feed.com*.
4.   Thomas, Ben. (November 3, 2015) "The Man Who Tried to Weigh the Soul." *Discover Magazine*.
5.   Smit, Jana Louise. (June 25, 2019) "10 Rare Recently Discovered Religious Artifacts." *Listverse.com*.
6.   Than, Ker. (April 30, 2010) "Noah's Ark Found in Turkey?" *National Geographic News*.

Chapter Sixteen: Insomnia

1.   Stroby, Wallace. (1991) "Writers on Writing." *Writer's Digest*.
2.   Mascetti, Gian Gastone. (July 12, 2016) "Unihemispheric Sleep and Asymmetrical Sleep: Behavioral, Neurophysiological, and Functional Perspectives." 2016:8: 221–238.
3.   Anderson, Scott Thomas. (2012) "Shadow People: How Meth-driven Crime Is Eating at the Heart of Rural America." *Coalition for Investigative Journalism*.

Chapter Seventeen: The Green Mile

1.   Argabrite, Angie. (March 17, 2000) "Studios Reveal Their Web Savvy in Sites for the Best Picture Nominees." *Entertainment Weekly*.

2. Balter, Michael. (June 11, 2012) "Aging is Recorded in Our Genes." *Science Mag.org.*

3. (1990) "Unproven Methods of Cancer Management: Psychic Surgery." *CA: A Cancer Journal for Clinicians.*

4. (August 7, 1890) "Far Worse Than Hanging." *New York Times.*

5. King, Stephen. (1996) *The Green Mile.* New American Library: New York.

Chapter Eighteen: The Girl Who Loved Tom Gordon

1. Hendrix, Grady. (July 24, 2015) "The Great Stephen King Reread: The Girl Who Loved Tom Gordon." *Tor.com.*

2. Urbigkit, Cat. (August 26, 2019) "Bear Attacks Increasing Worldwide." *Cowboy State Daily,*

3. Butler, Natalie. (January 17, 2018) "How Long Can You Live Without Food or Water?" *Healthline.*

4. Rajapaksa, Roshini. (January 16, 2017) "Can You Catch a Cold From Going Outside Without a Jacket?" *Health.com.*

5. (2018) "Top 20 Pneumonia Facts." *American Thoracic Society.*

6. (2018) Top 20 Pneumonia Facts." *American Thoracic Society.*

Chapter Nineteen: Gwendy's Button Box

1. Diaconis, Persi. (December 1989) "Methods for Studying Coincidences." *1989 American Statistical Association Journal of the American Statistical Association.*

2. Beck, Julie. (February 23, 2016) "Coincidences and the Meaning of Life." *The Atlantic.*

Chapter Twenty: Dreamcatcher

1. Greene, Andy. (October 31, 2014) "Stephen King: The Rolling Stone Interview." *Rolling Stone.*

2. Costa, Cheryl. (August 4, 2017) "UFOs, Aliens and Abductions Survey." *Syracuse New Times.*

3. McNamara, P., Dietrich-Egensteiner, L., & Teed, B. (2017) "Mutual Dreaming." *Dreaming.*

4. Roach, John. (2010) "The Real Science Behind Dream Research." *NBC News.com.*

5. Anderson, G. L. (2004) "Fungus Spore." *Science Direct.com.*

6. Payne, Kenneth Wilcox. (1928) "Is Telepathy all Bunk?" *Popular Science Monthly.*

7. Wild, Sarah. (May 26, 2018) "Are Viruses the New Frontier for Astrobiology?" *Space.com.*

Chapter Twenty-One: From a Buick 8

1. Miller, Laura. (September 29, 2002) "Not Your Father's Roadmaster." *The New York Times.*

2. (October 24, 2019) "Famous Cursed and Haunted Cars." *Volvo Cars.com.*
3. Marshall, Aarian. (March 26, 2020) "New Rules Could Finally Clear the Way for Self-Driving Cars." *Wired.*
4. McCausland, Phil. (November 9, 2019) "Self-driving Uber car that hit and killed woman did not recognize that pedestrians jaywalk." *NBC News.com.*
5. (June 30, 2016) "A Tragic Loss." *Tesla.com.*
6. Borenstein, Jason. Et al. (April 2019) "Self-Driving Cars and Engineering Ethics: The Need for a System Level Analysis." *Science & Engineering Ethics.*
7. Greene, Andy. (October 31, 2014) "Stephen King: The Rolling Stone Interview." *Rolling Stone.*

Chapter Twenty-Two: Cell

1. (2020) stephenking.com.
2. McVeigh, Daniel P. (2013) "An Early History of the Telephone: 1664–1866: Robert Hooke's Acoustic Experiments and Acoustic Inventions." *Columbia University.*
3. (October 23, 2018) "History of Mobile Cell Phones." *Be Businessed.com.*
4. Skaggs, Dr. Reed. (October-December 2007) "Exploiting Technical Opportunities to Capture Advanced Capabilities for Our Soldiers." *Army AL&T.*
5. Niiler, Eric. (May 25, 2018) "Sonic Weapons' Long, Noisy History." *History.com.*
6. (2020) "Is It a Good Idea to Store Batteries in a Refrigerator or Freezer?" *Energizer.com.*
7. Sharma, VP. et al. (2010) "Cell Phone Radiations Affect Early Growth of Vigna Radiata (mung bean) Through Biochemical Alterations." *Zeitschrift für Naturforschung.*
8. (2020) "8 Surprising Cell Phone Statistics." *Mobile Coach.com.*
9. Friederici, Peter. (March 2009) "How a Flock of Birds Can Fly and Move Together." *Audubon Magazine.*

Chapter Twenty-Three: Lisey's Story

1. MacDonald, Jay. (October 2006) "Stephen King Near Miss." *Book Page.*
2. (2012) "What is Dissociation and What to do About It?" *Washington.edu.*

Chapter Twenty-Four: Duma Key

1. Minzesheimer, Bob. (January 23, 2008) "'Duma Key' Finds Stephen King Stepping into His Own Life." *USA Today.*
2. Marshall, Barbara. (April 19, 2017) "Adios, Snowbirds: Glad Palm Beach County's Tourist Season is Over?" *The Palm Beach Post.*
3. Ziegler-Graham K. (2008) "Estimating the Prevalence of Limb Loss in the United States: 2005 to 2050." *Archives of Physical Medicine and Rehabilitation.*

4. Kelly, Patrick. (May 2011) "Survey of Phantom Limb Pain, Phantom Sensation and Stump Pain in Cambodian and New Zealand Amputees." *Pain Medicine.*

5. Harlow, John Martyn. (1868. "Recovery from the Passage of an Iron Bar through the Head." *Publications of the Massachusetts Medical Society.*

6. (2020) "Traumatic Brain Injuries." *CDC.gov.*

7. Gregoire, Carolyn. (January 14, 2014) "The Surprising Ways The Weather Affects Your Health And Well-Being." *The Huffington Post.*

8. (January 8, 2018) "Winter is the Season for Special Wound Care." *Advanced Tissue.com.*

9. Howard, Robert. (1992) "Folie a Deux Involving a Dog." *American Journal of Psychiatry.*

10. Kubler-Ross, Elisabeth. (1969) *On Death and Dying.* The MacMillan Company: New York.

11. Bonanno, George. (2009) "The Other Side of Sadness." *American Psychological Association.*

Chapter Twenty-Five: Under the Dome

1. Lileks, James. (November 17, 2009) "Self-proclaimed 'lazy' author Stephen King releases his 51st novel." *PopMatters.*

2. Archer, Cristina L. (June 2019) "Global Warming Will Aggravate Ozone Pollution in the U.S. Mid-Atlantic." *Journal of Applied Meteorology & Climatology.*

3. Schulze, Ernst Detlef. (March 2020) "The climate change mitigation effect of bioenergy from sustainably managed forests in Central Europe." *GCB Bioenergy.*

4. Gowthaman, Nirandi. (September 29, 2019) "12 motivational quotes by Greta Thunberg that will inspire you to change for the planet." *Your Story. com.*

5. Albeck-Ripka, Livia. (2020) "How to Reduce Your Carbon Footprint." *The New York Times.*

Chapter Twenty-Six: 11/22/63

1. McGinty, Stephen. (November 5, 2011) "Exclusive interview: Stephen King—the best-selling author speaks about his life, career and Scottish weather." *The Scotsman.*

2. (January 27, 2007) "Stephen King's Dark Tower." *Marvel Spotlight #14.*

3. (August 14, 2017) "The Butterfly Effect: Everything You Need to Know About This Powerful Mental Model." *Farnam Street.*

4. Greenspan, Jesse. (April 25, 2016) "8 Things You May Not Know About Chernobyl." *History.com.*

5. Spiegelman, Clifford. (December 7, 2017) "What Better Forensic Science Can Reveal About the JFK Assassination." *The Conversation.*

6.   Nalli, N. R. (2018). "Gunshot-wound Dynamics Model for John F. Kennedy Assassination." *Heliyon*.
7.   Andrews, Evan. (November 18, 2013) "9 Things You May Not Know About the Warren Commission." *History.com*.

Chapter Twenty-Seven: Doctor Sleep

1.   Sopp, Brian. (August 14, 2006) "Places of Our Dreams." *U.S. News & World Report*.
2.   (November 13, 2009) "Hungarian countesses' torturous escapades are exposed."*History.com*.
3.   (February 24, 2020) "What is Astral Projection?" *Gaia.com*.
4.   (February 24, 2020) "What is Astral Projection?" *Gaia.com*.

Chapter Twenty-Eight: Mr. Mercedes

1.   Treimer, Margaret. (August 1, 1988) "Subliminal Messages, Persuasion and Behavior Change." *Journal of Social Psychology*.
2.   Hauke Egermann, Reinhard. (December 1, 2006) *Journal of Articles in Support of the Null Hypothesis*.

Chapter Twenty-Nine: The Outsider

1.   (May 22, 2018) "Stephen King on 'The Outsider.'" *CBS This Morning*.
2.   (May 22, 2018) "Stephen King on 'The Outsider.'" *CBS This Morning*.
3.   Miller, Greg. (May 2010) "How Our Brains Make Memories." *Smithsonian Magazine*.
4.   (2020) "Eyewitness Identification Reform." *The Innocence Project.org*.
5.   Bailey, Melanie. (April 27, 2018) "The Hidden Data in Your Fingerprints." *Scientific American*.
6.   Oke, Chris. (July 23, 2016) "What the Dog Smelled: The Science and Mystery of Cadaver Dogs." *CBC News,*
7.   Schiro, George. (2020) "Collection and Preservation of Blood Evidence from Crime Scenes." *Crime Investigator Network.net*.
8.   (February 2020) "Holly Gibney." *Stephen King Wiki*.
9.   Phillips, Melissa Lee. (2020) "Synesthesia." *Washington.edu*.
10. Palmeri, Thomas J. (September 11, 2006) "What is Synesthesia?" *Scientific American*.

Chapter Thirty: The Institute

1.   Breznican, Anthony. (September 3, 2019) "Life Is Imitating Stephen King's Art, and That Scares Him." *The New York Times*.
2.   Brooks, Xan. (September 7, 2019) "Stephen King: 'I have outlived most of my critics. It gives me great pleasure.'" *The Guardian*.
3.   (June 16, 2017) "Mk-Ultra." *History.com*.

4. Richardson, Lindsay. (May 20, 2018) "Their lives were ruined: Families of MK Ultra survivors planning class-action suit." *CTV News.*
5. Radin, Dean. (March 18, 2018) "ESP: Inside the government's secret program of psychic spies." *CBS News.*

Index